MyMathLab®
Notebook

Gex, Incorporated

Basic Mathematics

John Squires
Chattanooga State Community College

Karen Wyrick
Cleveland State

Reproduced by Pearson from electronic files supplied by the author.

Publishing as Pearson, 75 Arlington Street, Boston, MA 02116.

Bound ISBN-13: 978-0-321-72547-9
Bound ISBN-10: 0-321-72547-6

Loose leaf ISBN-13: 978-0-321-75973-3
Loose leaf ISBN-10: 0-321-75973-7

1 2 3 4 5 6 BRR 15 14 13 12 11

MyMathLab Notebook

to accompany

MyMathLab Basic Mathematics
John Squires, Karen Wyrick

Table of Contents

Name: ______________________________ Date: ________________
Instructor: ____________________________ Section: ______________

Whole Numbers
1.1 Whole Numbers

Vocabulary
whole number • standard form • place value • expanded notation

1. To better understand the value of each digit in a number, we can write it in _____________.

Step-by-Step Video Notes
Watch the Step-by-Step Video lesson and complete the examples below.

Example	Notes
1. Consider the number 13,579. What digit is the tens place value? 7 What is the place value of the 3? ___________ What is the actual value of the 3? 3000	
2. Write 4295 in expanded form. The 4 is in the thousands place value, so the actual value is 4000. What is the actual value of the 2? ☐ What is the actual value of the 9? ☐ What is the actual value of the 5? ☐ $4295 = 4000 + \square 00 + \square 0 + 5$ Answer:	

Example	Notes
3. Convert 1209 to words. Write the number in the first period on the left followed by the period name and a comma. one ______________, Do this for the next period, but remember that the "ones" do not need the name of the period. Answer:	
4. Write twenty-four thousand, seven hundred twelve in standard form. Read the number from left to right. Write the number in the first period followed by a comma. ☐, Continue. Answer:	

Helpful Hints

When converting numbers to words or writing numbers in standard form, you always start at the left.

The word "and" is not included when reading or writing numbers.

Concept Check

1. How many place values are in a period?

Practice

Convert to words.

2. 5287

3. 321,608

Write in standard form.

4. two thousand six hundred four

5. twelve thousand, forty-eight

Name: ______________________________ Date: ________________
Instructor: ___________________________ Section: ______________

Whole Numbers
1.2 Rounding

Vocabulary
estimating • rounding • rounding down • rounding up

1. ____________ is finding a number close to the exact number, but easier to work with.

Step-by-Step Video Notes
Watch the Step-by-Step Video lesson and complete the examples below.

Example	Notes
1. Round 506,243 to the nearest thousand. Underline the digit in the place value to which you are rounding. 50<u>6</u>,243 What is the digit to the right of the underlined digit? □ It is 4 or less, so the underlined digit stays the same. Replace the digit to the right of 6 with zeros. Answer: 506,□□□	
2. Round 101,697 to the nearest hundred. 101,<u>6</u>97 Answer:	

Example	Notes
3. Round 1296 to the nearest ten. $12\underline{9}6$ Look at □, which is 5 or more so the under Lined digit increases by one. Note that when a 9 becomes a 10, the digit to the left of 9 also increases by one, so the 2 becomes a three. Answer: 13□□	
4. Round 449,985 to the nearest hundred. Answer:	

Helpful Hints

Remember to replace all the digits to the right of the digit you are rounding with zeros after rounding up or rounding down.

When a 9 is rounded up to ten, the digit to the left of the 9 must also be increased by one.

Concept Check

1. When rounding a number to the nearest place value, and the number to the right of the digit being rounded is 5, do you round up or round down?

Practice

Round the following numbers to the nearest place value indicated.

2. 4528 to the nearest ten
3. 23,179 to the nearest thousand
4. 894 to the nearest hundred
5. 49,926 to the nearest hundred

Name: ______________________________ Date: ________________

Instructor: ___________________________ Section: ______________

Whole Numbers
1.3 Adding Whole Numbers; Estimation

Vocabulary

addition • sum • estimation • commutative property of addition • addition property of zero • rounding • associative property of addition

1. ____________ occurs when you combine numbers.

2. The property which states that changing the order when adding numbers does not change the sum is the ____________.

Step-by-Step Video Notes

Watch the Step-by-Step Video lesson and complete the examples below.

Example	Notes
1. Find the sum of $5+6$. The sum is the answer to the addition, which is combining 5 and 6 and counting the total number of items. Answer:	
2. The statement $(3+5)+7=3+(5+7)$ is true. This shows that changing the grouping when adding numbers does not change the sum. Which property states this? Answer:	
3. Find the sum of $126+12$. Stack the numbers so that the digits in the same place value are lined up vertically. Begin at the right adding digits. $\begin{array}{r}126\\ +12\\ \hline\end{array}$ Answer:	

Example	Notes
4. Estimate by rounding to the nearest hundred and comparing to the exact answer. $7026 + 13,479$ Round 7026 to the nearest hundred. 7☐00 Round 13,479 to the nearest hundred. 13,☐00 Find the sum of the rounded numbers. ☐ What is the exact answer? ☐ Answer:	

Helpful Hints
When adding numbers vertically, if the sum of any column is more than 9, you need to carry the tens place value number to the next column.

Estimation is a process that can help you check that your exact answer is close to the actual answer.

Concept Check
1. Can you name three words or phrases that indicate addition?

Practice
Find the following.

2. $135 + 22$

3. $111 + 246 + 3189$

4. 13 more than 8

5. sum of 24 and 57

Name: ______________________________ Date: ______________
Instructor: ___________________________ Section: ______________

Whole Numbers
1.4 Subtracting Whole Numbers

Vocabulary

difference • sum • estimation • subtraction

1. ____________ of two numbers is taking away one number or quantity from another.

2. The ____________ is the answer to a subtraction problem.

Step-by-Step Video Notes

Watch the Step-by-Step Video lesson and complete the examples below.

Example	Notes
1. Subtract $10-4$. Take 4 away from 10. $10-4=\square$ Check by adding. $4+\square=10$ Answer:	
2. Subtract $57-38$. Begin with the ones place and subtract the bottom digit from the top. Since the top digit is smaller than the bottom digit, borrow from the next place value. $\begin{array}{r} 57 \\ \underline{-38} \end{array}$ Now subtract. $\begin{array}{r} 4\ 17 \\ \not{5}\ \not{7} \\ \underline{-3\ 8} \\ \square \end{array}$ Check by adding. Answer:	

Example	Notes
3. Subtract $232-141$. Check by adding. Answer:	
4. Estimate by rounding to the nearest thousand and comparing to the exact answer. $11,976-1245$ Round 11,976 to the nearest thousand. $\square$ Round 1245 to the nearest thousand. $\square$ Find the difference of the rounded numbers. Compare to the exact answer. Answer:	

Helpful Hints
Any number minus itself is zero.

Estimation is a process that can help you check that your exact answer is close.

Concept Check
1. Can you name three words or phrases that indicate subtraction?

Practice
Find the following.

2. $86-28$

3. $3045-2824$

4. 13 decreased by 8

5. 24 less than 52

Name: ______________________________ Date: ________________
Instructor: ___________________________ Section: ______________

Whole Numbers
1.5 Basic Problem Solving

Vocabulary

addition • perimeter • problem solving • translating

1. To find the ____________ of a figure, add the lengths of all its sides.

2. The procedure for ____________ involves creating a plan in symbols or words and performing calculations.

Step-by-Step Video Notes

Watch the Step-by-Step Video lesson and complete the examples below.

Example	Notes
1. Translate the sum of 5 and 3 to symbols. Enter the operation indicated by the word sum. 5☐3 Simplify. Answer:	
2. A rectangular garden measures 9 feet in length and is 5 feet wide. How many feet of fencing are needed to enclose the garden? Understand the problem. We are trying to find out how much ____________ is needed for the garden. Create a plan. We need to find the _________ of the lengths of the four sides. Find the answer. Check. Answer:	

Example	Notes
3. A local community college requires 64 credit hours for an Associate Degree. Kylie earned 28 credits this past year. How many more credit hours does Kylie need in order to get her degree? Enter the operation indicated from the key word or phrase in the problem. credit hours required ☐ credit hours earned is credit hours needed Answer:	
4. On Monday, Alex opened a checking account With an initial deposit of $300. She bought groceries for $75, spent $25 on gas, and spent $20 on new clothes. How much money is in her account after these purchases? Answer:	

Helpful Hints

When solving a problem, there are key words or phrases which can be translated into an operation.

After obtaining an answer from your calculation, make sure to check that this answers the question which was asked in the problem.

Concept Check

1. What two steps need to be done before calculation in solving a problem?

Practice

Today is Dan's turn to bring water to soccer practice. Dan's mom put 24 water bottles in the cooler and his dad put 6 water bottles to the cooler. Dan's soccer team has 19 players, but 3 are not at practice today. Each player takes one water bottle at practice.

2. Water bottles in cooler before practice.
3. Players at practice today.
4. Water bottles taken at practice.
5. Water bottles left in cooler after practice.

Name: ______________________________ Date: ________________
Instructor: ___________________________ Section: ______________

Whole Numbers
1.6 Multiplying Whole Numbers

Vocabulary

product • factors • addition • multiplication property of one • multiplication commutative property of multiplication • multiplication property of zero multiplication property of zero • distributive property of multiplication over addition

1. The ____________ is the answer to a multiplication problem.

2. The ____________ states that the changing the grouping when multiplying numbers does not change the product.

Step-by-Step Video Notes

Watch the Step-by-Step Video lesson and complete the examples below.

Example	Notes
1. For 3×4, identify the factors. 3, $\square$ Find the product. $3\times4=3+3+3+3=\square$ Answer:	
2. Rewrite $8(3+6)$ using the Distributive Property. $8(3)+\square(6)$ Simplify. Answer:	
3. Which property tells us that the following is true? $6\cdot25=25\cdot6$ Commutative Property, Associative Property, Distributive Property, Multiplication Property of Zero, Multiplication Property of One Answer:	

Example	Notes
4. Find the product $3 \cdot 248$. Multiply the bottom number by each digit on the top starting on the right. $3 \times 8 = 24$ The 4 is written as the ones digit answer and the 2 is carried to the next place value. $\begin{array}{r} ^{2} \\ 248 \\ \times 3 \\ \hline 4 \end{array}$ Complete the multiplication. Answer:	

Helpful Hints

The Distributive Property can be used to make mental calculations easier.

When number is carried to the next place value in a multiplication problem, this number is added to the product of the next multiplication.

Concept Check

1. Can you name two words or phrases that indicate multiplication?

Practice

Rewrite using the Distributive Property. Simplify.

2. $3(9+7)$

3. $7(5+2)$

Find the product.

4. $4 \cdot 164$

5. $3 \cdot 192$

Name: ______________________________ Date: ________________
Instructor: ___________________________ Section: ______________

Whole Numbers
1.7 Dividing Whole Numbers

Vocabulary
division • quotient • dividend • divisor • long division • remainder • divides exactly

1. The number you are dividing by is the ______________.

2. The ______________ is the answer to a division problem.

Step-by-Step Video Notes
Watch the Step-by-Step Video lesson and complete the examples below.

Example	Notes
1. Divide the following. $36 \div 4$ $4 \cdot \square = 36$, so $36 \div 4 = \square$ Answer:	
2. Divide the following. Use long division. $92 \div 4$ $\begin{array}{r} 2\,\square \\ 4\overline{)\,9\;\;2} \\ -8\;\;\; \\ \hline \square\,2 \\ -\square\,2 \\ \hline 0 \end{array}$ Answer:	

Example	Notes
3. Divide the following. Use long division. $92 \div 4$ $\square\square$ r $\square$ $3\overline{)1\ 4\ 8}$ $-\square\square$ $\square 8$ $-\square\square$ $\square$ Answer:	
4. Find the quotient of 21 and 7. $7 \cdot \square = 21$, so $21 \div 7 = \square$ Answer:	

Helpful Hints
In order to divide, you need to have mastered your multiplication facts. You can check any division problem using multiplication.

Remember that any non-zero number divided by itself is 1, and any number divided by 1 is the number itself. Zero divided by any non-zero number is 0, and division by 0 is undefined.

When translating for division, be careful to write the numbers in the correct order.

Concept Check
1. When using the long division symbol, which number goes inside? Which goes outside?

Practice
Divide the following.
2. $48 \div 6$

3. $119 \div 4$

Find the quotient of the following.
4. 32 and 8

5. 84 and 12

Name: ______________________________ Date: ______________

Instructor: ____________________________ Section: ______________

Whole Numbers
1.8 More with Multiplying and Dividing

Vocabulary

division • quotient • estimation • remainder • product • multiplication

1. The ______________ is the answer to a multiplication problem.

Step-by-Step Video Notes

Watch the Step-by-Step Video lesson and complete the examples below.

Example	Notes
1. Calculate the following product. $32 \cdot 125$ Set up the multiplication. It is preferable to put the larger number on top when setting up this problem. $\begin{array}{r} 1\ 2\ 5 \\ \times\ \ \ 3\ 2 \\ \hline \square\ 5\ 0 \\ \square\square\square\ 0 \\ \hline \square\square\square\ 0 \end{array}$ Answer:	
2. Multiply. 260(400) $26 \cdot 4 = \square$ Attach the 3 ending zeros from the factors to the end of the product. $260(400) = \square,\square\square\square$ Answer:	

Example	Notes
3. Estimate by rounding to the nearest ten. $7227(87) \approx 7230(\square)$ $723 \cdot \square = \square$ Attach the 2 ending zeros from the factors to the end of the product $260(400) \approx \square\square\square,\square\square\square$	
4. Divide the following. Set up the division. $1482 \div 12$ $\square\square\square$ r $\square$ 12) 1 4 8 2 $-\square\square$ $\square$ 8 $-\square\square$ $\square$ 2 $-\square\square$ $\square$	

Helpful Hints

When multiplying or dividing whole numbers with several digits, be careful to neatly stack the numbers vertically so that digits in the same place value line up.

Concept Check

1. When multiplying numbers ending in zeros, do you have to write each entire number in columns before multiplying? What can you do to make the multiplication simpler?

Practice

Calculate the following products.

2. $31 \cdot 19$

3. $24 \cdot 225$

Divide the following.

4. $1768 \div 12$

5. $1876 \div 16$

Name: ______________________________ Date: ________________
Instructor: ___________________________ Section: ______________

Whole Numbers
1.9 Exponents

Vocabulary
exponent • base • squared • cubed • power of 1 • power of 0

1. A(n) ____________ is a shortcut for repeated multiplication.

2. Any non-zero number raised to the ____________ is equal to 1.

Step-by-Step Video Notes
Watch the Step-by-Step Video lesson and complete the examples below.

Example	Notes
1. Identify the base and exponent, then evaluate. 6^2 The base is $\square$. The exponent is $\square$. $6^2 = 6 \cdot 6 = \square$ Answer:	
2. Write the following using exponents, then evaluate. $(3)(3)(3)$ $(3)(3)(3) = 3^{\square} = \square$ Answer:	
3. Evaluate. 31^0 Answer:	

Example	Notes
4. Calculate 10^2, 10^3 and 10^6. $10^2 = \square$ $10^3 = \square$ $10^6 = \square$	

Helpful Hints

The base is the number, or factor, being multiplied by itself. The exponent is the number of times the base is used as a factor.

Any number raised to the power of 1 is the number itself. Any non-zero number raised to the power of 0 is equal to 1.

An exponent expression, for example, 7^2 does NOT mean $7 \cdot 2$. It means use 7 as a factor 2 times, in other words, $7 \cdot 7$. The value of the expression is 49.

Concept Check

1. Can you explain a simple rule for evaluating powers of ten? How does the exponent relate to the number of zeros in the standard form number? Give an example.

Practice

Identify the base and exponent, then evaluate.

2. 4^3

3. 9^2

Evaluate.

4. 10^7

5. 25^0

Name: ______________________________ Date: ______________
Instructor: ____________________________ Section: ______________

Whole Numbers
1.10 Order of Operations and Whole Numbers

Vocabulary

order of operations • PEMDAS • parentheses • exponents

1. When simplifying an expression using the order of operations, always evaluate what is inside ____________ or other grouping symbols first.

Step-by-Step Video Notes

Watch the Step-by-Step Video lesson and complete the examples below.

Example	Notes
1. Calculate. $4+6\cdot 3$ The operations in this expression are addition and multiplication. Multiply first, and then add. $4+\square=\square$ Answer:	
2. Simplify. $3(4-2)^2+5$ First simplify the operation in the ____________, and then square that number. Next ____________ by $\square$, then ____________ $\square$. $3(\square)^2+5=3\cdot\square+5$ $=\square+5$ $=\square$ Answer:	

Example	Notes
3. Simplify by using the order of operations. $3^2 + 5 \cdot 4$ There is an exponent in the expression and multiplication. Evaluate the exponent then multiply. $\square + 5 \cdot 4 = \square + \square = \square$ Answer:	
4. Simplify. $\frac{22+10}{4 \cdot 5 - 4}$ Simplify the numerator. $\frac{\square}{4 \cdot 5 - 4}$ Simplify the denominator. $\frac{\square}{\square}$ Divide. $\square$	

Helpful Hints

These memory tips might help you remember the order of operations. P E MD AS, and Please Excuse My Dear Aunt Sally.

Other symbols also act like parentheses: brackets such as [ ] and { }, and fraction bars. When you see a fraction bar, act like there are parentheses around the numerator and denominator.

Concept Check

1. The rules for order of operations tell you to multiply and divide from left to right. Does the answer change if you multiply first if simplifying an expression such as $6 \div 3 \cdot 2$?

Practice

Evaluate.

2. $10 - 3 \cdot 2$

3. $4(9-6)^2 - 8$

Simplify.

4. $5^2 + 9 \div 3$

5. $\frac{14+13}{6 \cdot 2 - 3}$

Name: ______________________________ Date: ______________

Instructor: ____________________________ Section: ______________

Whole Numbers
1.11 More Problem Solving

Vocabulary

area • perimeter • problem solving • translating

1. The ____________ of a rectangle is found by multiplying its length times its width.

Step-by-Step Video Notes

Watch the Step-by-Step Video lesson and complete the examples below.

Example	Notes
1. Translate twice the sum of 7 and 4 to symbols. Enter the appropriate operation symbol. $2(7\square 4)$ Simplify. Answer:	
2. Find the area of a room that has a length of 21 feet and a width of 14 feet. Understand the problem. We need to find the ____________ of the room. Create a plan. We will find the ____________ by multiplying the length $\square$ feet by the width $\square$ feet. Find the answer. Check. Answer:	

Example	Notes
3. Desmond makes a salary of $39,000 per year. What is his monthly salary? We need to find _____________. There are ☐ months in a year. We will ☐ his yearly salary by ☐. Answer:	
4. The phone company charges $20 for installation and $25 for basic service. How much will Monique pay for one year of service with installation? Answer:	

Helpful Hints

When solving a problem, there are key words or phrases which can be translated into an operation.

After obtaining an answer from your calculation, make sure to check that this answers the question which was asked in the problem.

Concept Check

1. What two steps need to be done before calculation in solving a problem?

Practice

Find the area of the room with the following dimensions.

2. length of 21 feet and width of 12 feet

3. length of 10 feet and width of 8 feet

Joe earns $900 at his summer job where he works for three weeks, 20 hours each week.

4. How much did he earn each week?

5. How much did he earn per hour?

Name: ______________________________ Date: ________________
Instructor: ___________________________ Section: ______________

Factors and Fractions
2.1 Factors

Vocabulary
factor • divisibility rule(s) • common factor • greatest common factor (GCF)

1. A ______________ is a number that divides exactly into two or more numbers.

Step-by-Step Video Notes
Watch the Step-by-Step Video lesson and complete the examples below.

Example	Notes
1. Find the greatest common factor, or GCF, of 15 and 24. List the factors of 15. 1, □, □, 15 List the factors of 24. 1, 2, □, □, □, 8, □, □ Identify the factors common to both lists. 1, □ Choose the largest of these. □ Answer:	
2. Find the GCF of 18 and 54. List the factors of 18. 1, 2, □, □, □, □ List the factors of 54. 1, 2, □, □, □, □, □, □ Identify the factors common to both lists. Choose the largest of these. Answer:	

Example	Notes
3. Find the GCF of 132 and 198.	
4. List three ways to factor 18. List the factors of 18. 1, 2, □, □, □, □ Pair two factors that multiply to equal 18. $1 \cdot \square = 18$ Pair two different factors that equal 18. $2 \cdot \square = 18$ Repeat a third time. $\square \cdot \square = 18$	

Helpful Hints

There can be one or more common factors between two or more numbers, but there is only one greatest common factor (GCF).

The word "factor" can be either a noun or verb: Factors are the numbers being multiplied, and to factor a number means to write it as a product.

Concept Check

1. How many common factors are there between 24 and 96? Which is the GCF?

Practice

2. List all factors of 36.

3. Find the GCF of 30 and 85.

4. List three ways to factor 99.

Name: ______________________________ Date: ________________
Instructor: ____________________________ Section: ______________

Factors and Fractions
2.2 Prime Factorization

Vocabulary
factor • prime number • composite number • whole number

1. A ______________ is a whole number greater than 1 that has exactly two factors, the number itself and 1.

Step-by-Step Video Notes
Watch the Step-by-Step Video lesson and complete the examples below.

Example	**Notes**
1. Find the factors of 36. 1, 2, 3, 4, 6, ☐, ☐, ☐, 36	
2. Find the prime factorization of 24. Write 24 as a product of two factors. $24 = 2 \cdot \square$ Circle the factor(s) above that are prime numbers. Continue by writing 12 as a product of two factors. $12 = \square \cdot 4$ Circle the factor(s) above that are prime numbers. Continue by writing 4 as a product of two factors. $4 = 2 \cdot \square$ Circle the factor(s) above that are prime numbers. Write the prime factorization of 24 as a product of all circled prime numbers; repeated factors are written with exponents. Answer:	

Example	Notes
3. Find the prime factorization of 60.	

Helpful Hints

There is more than one way to write the factorization of a composite number. However, if all factors are rewritten as products of prime numbers, there is only one answer.

In the first step of rewriting a composite number as a product of two factors, there is often more than one combination of two factors that can be selected. However, continuing the steps of prime factorization will lead to the same final answer.

Concept Check

1. In finding the prime factorization of 72, if 72 is written as $6 \cdot 12$, what is the following step? And the next step?

Practice

2. Find the prime factorization of 56.

3. Find the prime factorization of 98.

4. Find the prime factorization of 210.

Name: ______________________________ Date: ________________

Instructor: ____________________________ Section: ______________

Factors and Fractions
2.3 Understanding Fractions

Vocabulary

fraction • denominator • undefined • improper fraction

1. A(n) ______________ is the bottom number in a fraction and indicates the number of parts in the whole.

2. A(n) ______________ is a fraction where the numerator is greater than or equal to the denominator.

Step-by-Step Video Notes

Watch the Step-by-Step Video lesson and complete the examples below.

Example	Notes
1. How many parts are shaded in the diagram? How many total parts are in the whole diagram? □ Write a fraction that represents the shaded part. Answer:	
2. Write a fraction that represents the shaded part.	

Example	Notes
3. The numerator of $\frac{4}{9}$ is ___ the denominator. Identify this fraction as proper or improper. Answer:	
4. Identify $\frac{11}{4}$ as proper or improper. Answer:	

Helpful Hints

When the numerator is equal to the denominator of a fraction, this is an improper fraction that is equal to 1.

A fraction with a numerator of zero is equal to zero; a fraction with a denominator of zero is undefined.

Concept Check

1. There are 9 boys in the class of 20 students. Represent this as a fraction and state why this is a proper fraction.

Practice

2. Write a fraction that represents the shaded part.

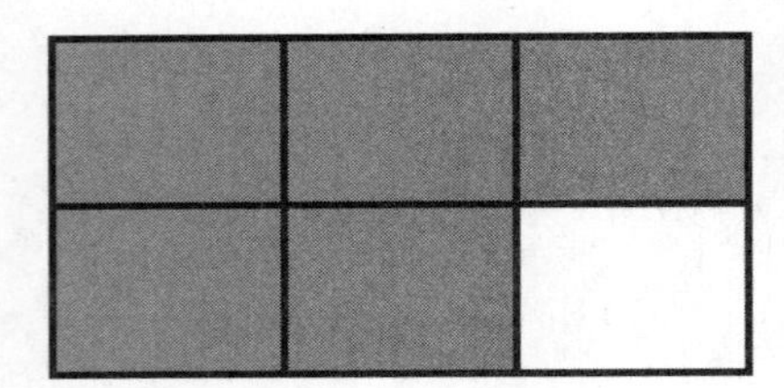

3. Identify $\frac{7}{7}$ as proper or improper.

4. Identify $\frac{2}{9}$ as proper or improper.

5. Identify $\frac{0}{5}$ as proper or improper.

Name: ______________________________ Date: ________________
Instructor: ____________________________ Section: ______________

Factors and Fractions
2.4 Simplifying Fractions – GCF and Factors Method

Vocabulary
equivalent fractions • factors • simplest form of a fraction • equivalent form of a fraction

1. ______________ are fractions that represent the same value.

2. A fraction is said to be in ______________ if there is no common factor other than 1 that divides exactly into the numerator and the denominator.

Step-by-Step Video Notes
Watch the Step-by-Step Video lesson and complete the examples below.

Example	Notes
1. Write $\frac{9}{15}$ in simplest form using the GCF method. Find the GCF of 9 and 15. ☐ Divide the numerator and the denominator by the GCF. Answer:	
2. Write $\frac{72}{96}$ in simplest form using the GCF method.	

Example	Notes
3. Write $\frac{20}{32}$ in simplest form using the factors method. Name a common factor of 20 and 32. $\boxed{2}$ Divide numerator and denominator by 2. $\frac{\square}{16}$ Name a common factor of $\square$ and 16. $\square$ Divide the numerator and denominator by this common factor. Answer:	

Helpful Hints

Dividing the numerator and denominator by a common factor results in an equivalent fraction; however, this is not always the simplest form of the fraction. The procedure must be repeated until there is no common factor of the numerator and denominator except for 1.

Using the GCF method leads to a quicker solution process than the factors method; however, if you are having a hard time finding the GCF, then the factors method is helpful.

Concept Check

1. Which method would you use for simplifying $\frac{6}{9}$? For simplifying $\frac{48}{175}$? Describe the first step for each method selected for the above fraction simplifications.

Practice

2. Write $\frac{24}{27}$ in simplest form using the GCF method.

3. Write $\frac{35}{175}$ in simplest form using the factors method.

4. Write $\frac{96}{144}$ in simplest form using either the GCF method or the factors method.

Name: ______________________________ Date: ________________
Instructor: ___________________________ Section: ______________

Factors and Fractions
2.5 Simplifying Fractions – Prime Factors Method

Vocabulary
prime factors method • composite number • common factor • simplest form

1. The prime factor method for simplifying fractions involves writing the numerator and denominator as products of prime numbers, then dividing the numerator and denominator by the ______________.

Step-by-Step Video Notes
Watch the Step-by-Step Video lesson and complete the examples below.

Example	Notes
1. Write $\frac{24}{36}$ in simplest form using the prime factors method. Write the numerator as the product of prime numbers. $24 = 2^3 \cdot \square$ Write the denominator as the product of prime numbers. $36 = 2^2 \cdot \square$ Divide the numerator and the denominator by the common factors. Answer:	

Example	Notes
2. Write in simplest form using the prime factors method. $\frac{84}{91}$ Write the numerator as a product of prime numbers. □ Write the denominator as a product of prime numbers. □ Divide the numerator and the denominator by the GCF. Multiply the remaining factors. Answer:	
3. Write $\frac{39}{135}$ in simplest form using the prime factors method.	

Helpful Hints

When simplifying fractions using the prime factor method, you must make sure that both the numerator and the denominator are written as the product of only prime numbers.

If the numerator is a factor of the denominator, then the simplified fraction will have numerator of 1.

Concept Check

1. How is the prime factors method for simplifying fractions similar to the GCF method?

Practice

2. Write $\frac{24}{27}$ in simplest form using the prime factors method.

3. Write $\frac{35}{175}$ in simplest form using the prime factors method.

4. Write $\frac{20}{120}$ in simplest form using the prime factors method.

Name: ______________________________ Date: ________________
Instructor: ___________________________ Section: ______________

Factors and Fractions
2.6 Multiplying Fractions

Vocabulary
numerator • denominator • multiplying fractions • equivalent fraction

1. Multiplying fractions can be done by first multiplying the numerators to get the ____________ of the product and multiplying the denominators to get the ____________ of the product.

Step-by-Step Video Notes
Watch the Step-by-Step Video lesson and complete the examples below.

Example	Notes
1. Multiply. Write your answer in simplest form. $\frac{4}{7}\cdot\frac{3}{5}$ Multiply the numerators. $4\cdot 3=\square$ Multiply the denominators. $7\cdot\square=\square$ Simplify. Answer:	
2. Multiply. Write your answer in simplest form. $\frac{7}{10}\cdot\frac{5}{8}$ Answer:	

Example	Notes
3. Multiply $\frac{4}{5}\cdot\frac{9}{16}$ by simplifying first. Write as one fraction. Do not multiply yet. Divide common factors in the numerator and denominator. Multiply the remaining factors. Answer:	
4. Multiply $3\cdot\frac{2}{9}$ by simplifying first.	

Helpful Hints
Both the numerators and the denominators must be multiplied when multiplying fractions; fractions should be written in simplest form.

When fractions are being multiplied and there are common factors in one or more of the numerators with one or more of the denominators, it is easier to simplify the fractions by dividing the numerator and the denominator by the common factors before multiplying.

Concept Check

1. Why is $\frac{7}{100}\cdot\frac{10}{21}$ an easier multiplication problem than $\frac{5}{16}\cdot\frac{7}{11}$?

Practice

2. Multiply $\frac{8}{15}\cdot\frac{3}{5}$. Write your answer in simplest form.

3. Multiply $\frac{2}{9}\cdot\frac{3}{5}$ by simplifying first.

4. Multiply $7\cdot\frac{3}{28}$ by simplifying first.

Name: ______________________________ Date: ________________
Instructor: ___________________________ Section: ______________

Factors and Fractions
2.7 Dividing Fractions

Vocabulary
reciprocals • common factors

1. Two numbers are ______________ of each other if their product is 1.

Step-by-Step Video Notes
Watch the Step-by-Step Video lesson and complete the examples below.

Example	Notes
1. Find the reciprocal of $\frac{3}{4}$. The reciprocal is $\frac{4}{\square}$	
2. Find the reciprocal of 5.	
3. Divide $\frac{3}{5} \div \frac{1}{2}$. Write in simplest form. Take the reciprocal of the second fraction. $\frac{\square}{\square}$ Multiply the reciprocal by the first fraction. $\frac{3}{5} \cdot \frac{\square}{\square} = \frac{\square}{\square}$ Answer:	

Example	Notes
4. Divide $\frac{6}{7} \div \frac{3}{2}$. Write in simplest form. Answer:	

Helpful Hints

To find the reciprocal of a whole number, first write the number as a fraction by putting it over 1.

When fractions are being divided, the reciprocal of the second fraction must be taken before the fractions are multiplied.

The numerator and denominator can be divided by common factors when dividing fractions, but only after the division has been changed to multiplication by the inverted second fraction.

Concept Check

1. How can $\frac{2}{5} \div \frac{3}{7}$ be rewritten as a multiplication?

Practice

2. Divide $\frac{8}{15} \div \frac{3}{4}$. Write your answer in simplest form.

3. Divide $8 \div \frac{2}{5}$. Write your answer in simplest form.

4. Divide $\frac{2}{7} \div 6$. Write your answer in simplest form.

Name: ______________________________ Date: ________________
Instructor: ___________________________ Section: ______________

LCM and Fractions
3.1 Finding the LCM – List Method

Vocabulary
multiple • greatest common multiple • least common multiple • whole number

1. A _______________ of a number is the product of a number and a positive whole number.

2. The ______________ of two or more numbers is the smallest number that is a multiple of the given numbers.

Step-by-Step Video Notes
Watch the Step-by-Step Video lesson and complete the examples below.

Example	Notes
1. Find the first five multiples of 9. 9, 18, ☐, 36, 45	
2. Find the first four multiples of 20. 20, ☐, ☐, ☐	
3. Find the least common multiple of 9 and 12. List the first several multiples of 9. 9, 18, ☐, 36, 45, ☐, ☐, 72 List the first several multiples of 12. 12, 24, ☐, ☐, 60, 72 Identify the multiples common to each list. ☐, 72 Choose the least of these. Answer:	

Example	Notes
4. Find the least common multiple of 6, 8 and 12.	

Helpful Hints

We can find a common multiple of two numbers by multiplying them together, but this may or may not be the least common multiple.

Be careful not to confuse the GCF, the greatest common factor, with the LCM, the least common multiple; the GCF is the greatest factor that can be divided evenly into the given numbers, while the LCM is the smallest number that is a multiple of the given numbers.

The GCF of two numbers is always less than or equal to the given numbers, while the LCM is always greater than or equal to the given numbers.

Concept Check

1. The following statement is false: The LCM of 4 and 10 is 40 because $4 \cdot 10 = 40$. Why is this a false statement?

Practice

2. List the first five multiples of 16.

3. List the first five multiples of 20.

4. Find the least common multiple of 16 and 20.

5. Find the least common multiple of 7, 9, and 21.

Name: ______________________________ Date: ________________
Instructor: ____________________________ Section: ______________

LCM and Fractions
3.2 Finding the LCM – GCF Method

Vocabulary
multiply • greatest common factor • least common multiple • divide

1. When finding the LCM of two numbers using the GCF method, first find the GCF of the two numbers, then ______________ the two original numbers and divide by the GCF.

Step-by-Step Video Notes
Watch the Step-by-Step Video lesson and complete the examples below.

Example	Notes
1. Find the LCM of 6 and 9 using the GCF method. Find the GCF of 6 and 9. $\square$ Multiply 6 and 9. $6 \cdot 9 = \square$ Divide product by GCF. Answer:	
2. Find the LCM of 6 and 15 using the GCF method. Find the GCF of 6 and 15. $\square$ Multiply 6 and 15. $\square \cdot \square = \square$ Divide product by GCF. Answer:	

Example	Notes
3. Find the LCM of 4 and 18 using the GCF method.	
4. Find the LCM of 7 and 9 using the GCF method.	

Helpful Hints

If the GCF of two numbers is 1, then the LCM is the product of the two numbers.

Finding the LCM of two or more numbers using the GCF method can be quicker than listing out multiples of each number.

Concept Check

1. Listing multiples of 3 and 42 to determine the LCM leads to a long list for the number 3. List the steps needed for finding these two numbers using the GCF method.

Practice

2. Find the LCM of 6 and 42 using the GCF method.
3. Find the LCM of 10 and 25 using the GCF method.
4. Find the LCM of 8 and 20 using the GCF method.
5. Find the LCM of 5 and 11 using the GCF method.

Name: ______________________________ Date: ________________
Instructor: ____________________________ Section: ______________

LCM and Fractions
3.3 Finding the LCM – Prime Factor Method

Vocabulary
finding the LCM • common prime factors • prime factorization • multiples

1. The _____________ of any whole number is the factored form in which all factors are prime numbers.

Step-by-Step Video Notes
Watch the Step-by-Step Video lesson and complete the examples below.

Example	Notes
1. Find the LCM of 12 and 42 using the prime factor method. Find the prime factorization of 12. $12 = 2 \cdot 2 \cdot 3$ Find the prime factorization of 42. $42 = 2 \cdot 3 \cdot$ ___ Write the prime factorizations one below the other, putting the common prime factors below each other. (table below) Write down the prime factor from each column. (table below) Multiply the list of prime factors. Answer:	

2	2	3	
2		3	□

2	2	3	□

Example	Notes
2. Find the LCM of 8 and 30 using the prime factor method. Write the prime factorizations of 8 and 30 one below the other, putting the common prime factors below each other. Multiply the list of prime factors. Answer:	
3. Find the LCM of 13 and 19 using the prime factor method.	

Helpful Hints

When using the prime factor method to find the LCM of two numbers, make sure to enter a prime factor in a column for each time it is used. Do not enter a prime factor with an exponent.

When using the prime factor method to find the LCM of two numbers, make sure to enter a prime factor from each column, even if the prime factor is not a factor of both of the original numbers.

Concept Check

1. How many total prime factors (matching the total number of columns) are there when finding the LCM of 24 and 30?

Practice

2. Find the LCM of 3 and 42 using the prime factor method.

3. Find the LCM of 10 and 25 using the prime factor method.

4. Find the LCM of 15 and 14 using the prime factor method.

5. Find the LCM of 8 and 9 using the prime factor method.

Name: ____________________ Date: ____________

Instructor: ____________________ Section: ____________

LCM and Fractions
3.4 Writing Fractions with an LCD

Vocabulary

least common denominator (LCD) • prime numbers • common factor • equivalent fractions

1. The ____________ is the least common multiple (LCM) of the denominators.

2. Fractions are ____________ if they have the same value.

Step-by-Step Video Notes

Watch the Step-by-Step Video lesson and complete the examples below.

Example	Notes
1. Find the LCM of 4 and 12. ☐ Find the LCD of $\frac{3}{4}$ and $\frac{5}{12}$. Answer:	
2. Write a fraction equivalent to $\frac{1}{6}$ using denominator of 12. What number does 6 need to be multiplied by to result in 12? ☐ Multiply the numerator and denominator of the original fraction by the same number. $\frac{1}{6} \cdot \frac{\square}{\square} = \frac{\square}{12}$ Answer:	

Example	Notes
3. Rewrite $\frac{1}{6}$ and $\frac{7}{9}$ using the LCD as the denominator. Find the LCD. $\square$ Find the number needed to multiply the denominator by to get the LCD for $\frac{1}{6} \cdot \frac{\square}{\square}$ Do the same for $\frac{7}{9} \cdot \frac{\square}{\square}$ Rewrite each fraction as equivalents with the LCD as the denominator. Answer:	
4. Rewrite $\frac{4}{15}$ and $\frac{5}{6}$ using the LCD as the denominator.	

Helpful Hints

If one denominator divides exactly into the other, then the LCD is the larger number.

If two denominators have no common factor other than 1, then the LCD is the product of the two denominators.

Concept Check

1. When rewriting $\frac{1}{3}$ and $\frac{1}{5}$ using the LCD, how do you know that neither numerator will be 1?

Practice

2. Rewrite $\frac{2}{3}$ using 9 as the denominator.
3. Rewrite $\frac{5}{9}$ and $\frac{2}{15}$ using the LCD as the denominator.
4. Rewrite $\frac{1}{4}$ and $\frac{5}{6}$ using the LCD as the denominator.

Name: ______________________________ Date: ________________
Instructor: ___________________________ Section: ______________

LCM and Fractions
3.5 Adding and Subtracting Like Fractions

Vocabulary
like fractions • unlike fractions • numerators • equivalent fractions

1. Fractions with the same, or common, denominator are called ______________.

2. Fractions without a common denominator are called ______________.

Step-by-Step Video Notes
Watch the Step-by-Step Video lesson and complete the examples below.

Example	Notes
1. Add the like fractions $\frac{2}{5}+\frac{1}{5}$. Simplify if possible. Add the numerators. $2+1=\square$ Keep the denominator. $\square$ Answer:	
2. Add the like fractions $\frac{3}{10}+\frac{3}{10}$. Simplify if possible.	

Example	Notes
3. Subtract the like fractions $\frac{6}{7}-\frac{2}{7}$. Simplify if possible. Subtract the numerators. $6-2=\square$ Keep the denominator. $\square$ Answer:	
4. Subtract the like fractions $\frac{11}{14}-\frac{5}{14}$. Simplify if possible.	

Helpful Hints

When adding or subtracting like fractions, the operation is done on the numerators only.

When adding or subtracting like fractions, the denominator remains unchanged.

Concept Check

1. Whether adding or subtracting $\frac{4}{5}$ and $\frac{2}{5}$, what is the denominator of the resulting fraction? What will the numerator of the result of addition be? What will the numerator of the result of subtraction be?

Practice

2. Add the like fractions $\frac{4}{11}+\frac{3}{11}$. Simplify if possible.

3. Add the like fractions $\frac{1}{18}+\frac{2}{18}+\frac{11}{18}$. Simplify if possible.

4. Subtract the like fractions $\frac{5}{9}-\frac{1}{9}$. Simplify if possible.

5. Subtract the like fractions $\frac{5}{8}-\frac{3}{8}$. Simplify if possible.

Name: ______________________________ Date: ______________
Instructor: ___________________________ Section: ______________

LCM and Fractions
3.6 Adding and Subtracting Unlike Fractions

Vocabulary
like fractions • unlike fractions • prime factors • denominator

1. Fractions without a common denominator are called ____________.

Step-by-Step Video Notes
Watch the Step-by-Step Video lesson and complete the examples below.

Example	Notes
1. Add $\frac{3}{4}+\frac{1}{6}$. Simplify if possible. Find the LCD. 12 Rewrite as like fractions with the LCD of 12 as the denominator. $\frac{3}{4}=\frac{\square}{12}, \frac{1}{6}=\frac{\square}{12}$ Add the numerators. $\square$ Keep the denominator. $\square$ Answer:	
2. Add $\frac{3}{8}+\frac{2}{5}$. Simplify if possible.	

Example	**Notes**
3. Subtract $\frac{5}{6}-\frac{1}{3}$. Simplify if possible. Find the LCD. ☐ Rewrite the fractions as like fractions with the LCD as the denominator. Subtract the numerators. Keep the denominator. Answer:	
4. Subtract $\frac{7}{10}-\frac{4}{15}$. Simplify if possible.	

Helpful Hints
When rewriting fractions with a common denominator, the LCD is usually used.

When adding or subtracting unlike fractions, a common denominator is needed.

Concept Check
What important step(s) must be done before adding or subtracting unlike fractions?

Practice

1. Add $\frac{2}{3}+\frac{1}{5}$. Simplify if possible.
2. Add $\frac{5}{6}+\frac{1}{18}$. Simplify if possible.
3. Subtract $\frac{3}{4}-\frac{1}{10}$. Simplify if possible.
4. Subtract $\frac{5}{7}-\frac{3}{14}$. Simplify if possible.

Name: ______________________________ Date: ______________
Instructor: ___________________________ Section: ______________

LCM and Fractions
3.7 Order of Operations and Fractions

Vocabulary
order of operations • addition of fractions • multiplication of fractions • LCM

1. The procedure for the ______________ is to evaluate what is inside parentheses, evaluate any exponents, perform multiplication/division, and then perform addition/subtraction.

Step-by-Step Video Notes
Watch the Step-by-Step Video lesson and complete the examples below.

Example	Notes
1. Simplify $\left(\frac{1}{3}-\frac{1}{6}\right)+\frac{1}{2}$ by using the order of operations. Evaluate what is inside the parentheses. $\frac{1}{3}-\frac{1}{6}=\frac{\square}{\square}$ Now perform the addition. $\frac{\square}{\square}+\frac{1}{2}=\frac{\square}{\square}$ Answer:	
2. Simplify $\frac{1}{4}\div\frac{3}{2}+\frac{1}{2}\cdot\frac{1}{3}$ by using the order of operations. Perform the multiplication and division from left to right. $\frac{1}{4}\div\frac{3}{2}=\frac{\square}{\square}$, $\frac{1}{2}\cdot\frac{1}{3}=\frac{1}{6}$ Now perform the addition. $\frac{\square}{\square}+\frac{1}{6}=\frac{\square}{\square}$ Answer:	

Example	Notes
3. Simplify $\left(\frac{3}{4}\right)\left(\frac{1}{2}-\frac{1}{4}\right)^2+\frac{2}{5}\cdot\frac{1}{2}$ using the order of operations.	

Helpful Hints

PE MD AS is an acronym to help remember the order of operations. This can be remembered as Please Excuse My Dear Aunt Sally; the letter P stands for parentheses, E for exponents, MD for multiplication/division, and AS for addition/subtraction.

After addressing any parentheses and exponents, make sure to do the multiplication/division from left to right, then the addition/subtraction from left to right.

Concept Check

Why is the first step to simplifying $\frac{1}{5}+\frac{2}{5}\cdot\left(\frac{1}{4}\right)^2$ not adding $\frac{1}{5}+\frac{2}{5}$, nor multiplying $\frac{2}{5}\cdot\frac{1}{4}$? What is the second step? The third?

Practice

1. Simplify $\frac{1}{7}+\frac{2}{7}\cdot\left(\frac{1}{2}\right)^2$ using the order of operations.

2. Simplify $\frac{1}{2}-\frac{1}{3}\div\frac{5}{12}$ using the order of operations.

3. Simplify $\frac{1}{4}\cdot\left(\frac{3}{5}+\frac{1}{10}\right)\div\frac{3}{10}$ using the order of operations.

Name: ______________________________ Date: ______________

Instructor: ___________________________ Section: ______________

Mixed Numbers

4.1 Changing a Mixed Number to an Improper Fraction

Vocabulary

mixed number • improper fraction • numerator • whole numbers

1. A(n) ____________ is the sum of a whole number and a fraction.

Step-by-Step Video Notes

Watch the Step-by-Step Video lesson and complete the examples below.

Example	Notes
1. Write $2\frac{1}{3}$ as an improper fraction. Multiply the denominator by the whole number. $\square \times \square = 6$ Add this to the numerator. $6 + \square = \square$ Write this value over the original denominator. $\frac{\square}{6}$ Answer:	
2. Write $4\frac{5}{7}$ as an improper fraction. Multiply the denominator by the whole numerator. Add this to the numerator. Answer:	

Example	Notes
3. Exchange a \$5 bill for 20 quarters at a car wash. Place the whole number over 1. $\frac{5}{1}$ We want a denominator of 4, as there are 4 quarters in 1 dollar. Multiply the numerator and denominator by 4. $\frac{5\times4}{1\times\square}=\frac{\square}{\square}$ Answer:	
4. Write 91 as an improper fraction. Choose a denominator other than 1. Answer:	

Helpful Hints

When changing a mixed number to an improper fraction, the denominator will always be the same.

Whole numbers can be changed to improper fractions having any denominator desired; the numerator is multiplied by the chosen denominator.

Concept Check

1. When changing $2\frac{4}{5}$ to an improper fraction, explain why the numerator will be 14.

Practice

Write as an improper fraction.

2. $5\frac{9}{20}$

3. $10\frac{4}{7}$

4. 82

5. $12\frac{1}{6}$

Name: ______________________________ Date: ______________

Instructor: ____________________________ Section: ______________

Mixed Numbers
4.2 Changing an Improper Fraction to a Mixed Number

Vocabulary

remainder • quotient • numerator • denominator

1. When changing an improper fraction to a mixed number, the ____________ is the whole number part.

Step-by-Step Video Notes

Watch the Step-by-Step Video lesson and complete the examples below.

Example	Notes
1. Write $\frac{13}{5}$ as a division problem. $13 \div \square$ Answer:	
2. Write $\frac{13}{4}$ as a mixed number. Divide the numerator by the denominator. $4\overline{)13}$ with $\square$ above What is the remainder? $\square$ The quotient is the whole number part. The remainder is the numerator of the fractional part. The denominator stays the same. $\square\frac{\square}{4}$ Answer:	

Example	Notes
3. Write $\frac{7}{5}$ as a mixed number. Answer:	
4. Write $\frac{15}{5}$ as a mixed number. Answer:	

Helpful Hints

When setting up a division problem to change an improper fraction into a mixed number, the top number (the numerator) goes inside the long division sign.

If the remainder is zero when changing an improper fraction into a mixed number, then the answer is a whole number; if there is a remainder, this becomes the numerator of the mixed number.

Concept Check

1. When changing an improper fraction to a mixed number, what does the whole number represent?

Practice

Write each as a mixed number or whole number.

2. $\frac{11}{3}$

3. $\frac{17}{4}$

4. $\frac{24}{6}$

5. $\frac{25}{7}$

Name: ______________________________ Date: ______________

Instructor: ____________________________ Section: ______________

Mixed Numbers
4.3 Multiplying Mixed Numbers

Vocabulary

factor • mixed number • whole number • improper fraction

1. When multiplying mixed numbers, the first step is to rewrite each mixed number as a ____________.

Step-by-Step Video Notes

Watch the Step-by-Step Video lesson and complete the examples below.

Example	Notes
1. Multiply $1\frac{1}{2} \cdot 5\frac{1}{6}$. Write $1\frac{1}{2}$ and $5\frac{1}{6}$ as improper fractions. $1\frac{1}{2} = \frac{\square}{2}$ $\quad 5\frac{1}{6} = \frac{\square}{6}$ Multiply the fractions. $\frac{\square}{2} \cdot \frac{\square}{\square} = \frac{\square}{\square}$ Rewrite as a mixed or whole number. Answer:	
2. Multiply $1\frac{2}{3} \cdot 4\frac{1}{5}$. Write each mixed number as an improper fraction. $\frac{\square}{3} \cdot \frac{\square}{5}$ Multiply and rewrite answer as a mixed or whole number. Answer:	

Example	Notes
3. Multiply $2\left(3\frac{1}{2}\right)\left(1\frac{2}{7}\right)$. Answer:	

Helpful Hints

Remember when multiplying fractions, the numerators are multiplied and the denominators are multiplied.

Make sure to simplify the fraction answer in either the resulting improper fraction or the fraction of the mixed number.

Concept Check

1. When multiplying mixed numbers, what form of the numbers must be used?

Practice

Multiply.

2. $2\frac{2}{3}\cdot 1\frac{1}{4}$

4. $1\frac{5}{6}\cdot 2\frac{4}{7}$

3. $1\frac{2}{7}\cdot 2\frac{1}{3}$

5. $3\left(1\frac{2}{3}\right)\left(2\frac{4}{5}\right)$

Name: ______________________________ Date: ________________
Instructor: ___________________________ Section: ______________

Mixed Numbers
4.4 Dividing Mixed Numbers

Vocabulary
mixed number • whole number • factor • improper fraction

1. When dividing mixed numbers, each mixed number should be written as a(n) ____________.

Step-by-Step Video Notes
Watch the Step-by-Step Video lesson and complete the examples below.

Example	**Notes**
1. Divide. $1\frac{3}{4} \div 2\frac{4}{5}$	
Rewrite $1\frac{3}{4}$ as an improper fraction. $\frac{\square}{4}$	
Rewrite $2\frac{4}{5}$ as an improper fraction. $\frac{\square}{5}$	
Divide the fractions by inverting the second fraction and multiplying it by the first fraction.	
$\frac{\square}{4} \cdot \frac{\square}{\square} = \frac{\square}{\square}$	
Rewrite answer as a mixed or whole number.	
Answer:	

Example	Notes
2. Divide. $3\frac{2}{5} \div 2\frac{2}{3}$ Write each mixed number as an improper fraction. $\frac{\square}{5} \div \frac{\square}{3}$ Divide the fractions. Rewrite answer as a mixed or whole number. Answer:	
3. Divide. $4 \div 3\frac{1}{3}$ Answer:	

Helpful Hints

Remember that whole numbers can be rewritten as improper fractions by placing the original number as the numerator and making the denominator 1.

Always simplify a fraction by dividing both numerator and denominator by common factors.

Concept Check

1. Why must mixed numbers be changed to improper fractions before dividing them?

Practice

Divide.

2. $1\frac{1}{4} \div 3\frac{1}{3}$

3. $4\frac{2}{5} \div 2\frac{3}{4}$

4. $6 \div 2\frac{4}{5}$

5. $3\frac{5}{6} \div \frac{1}{6}$

Name: ______________________________ Date: ________________
Instructor: ____________________________ Section: ______________

Mixed Numbers
4.5 Adding Mixed Numbers

Vocabulary

mixed number • common denominator • numerator • denominator

1. In order to add fractions, they must have the same ____________.

Step-by-Step Video Notes

Watch the Step-by-Step Video lesson and complete the examples below.

Example	Notes
1. Add. $3\frac{1}{3}+2\frac{1}{3}$ Add the like fractions by adding the numerators and keeping the denominator. $\frac{1}{3}+\frac{1}{3}=\frac{\square}{3}$ Add the whole numbers. $3+2=\square$ Answer:	
2. Add. $7\frac{1}{3}+4\frac{3}{5}$ Rewrite the fractions with a common denominator. $\frac{1}{3}=\frac{\square}{\square}, \quad \frac{3}{5}=\frac{\square}{\square}$ Add the fractions. Add the whole numbers. Answer:	

Example	Notes
3. Add. $2\frac{5}{6}+9\frac{2}{3}$ Answer:	
4. Add. $2\frac{1}{3}+15\frac{2}{3}$ When adding fractions yields an improper fraction, it needs to be changed to a mixed or whole number. This is then added to the sum of the whole numbers. $\frac{1}{3}+\frac{2}{3}=\frac{\square}{3}=\square$ Answer:	

Helpful Hints
Fractions must have the same denominator to be added; fractions can be rewritten with a common denominator, the LCD.

Always simplify fractions.

Concept Check
1. What are the steps used when adding mixed numbers?

Practice
Add.

2. $4\frac{1}{5}+2\frac{2}{5}$

3. $2\frac{3}{4}+3\frac{1}{4}$

4. $1\frac{7}{10}+2\frac{4}{5}$

5. $5\frac{1}{2}+3\frac{1}{3}$

Name: ______________________________ Date: ________________

Instructor: ______________________________ Section: ______________

Mixed Numbers
4.6 Subtracting Mixed Numbers

Vocabulary

mixed number • denominator • numerator • improper fraction

1. In order to subtract fractions, they must have the same ____________.

Step-by-Step Video Notes

Watch the Step-by-Step Video lesson and complete the examples below.

Example	Notes
1. Subtract $5\frac{4}{7}-2\frac{1}{7}$. Subtract the fractions. $\frac{4}{7}-\frac{1}{7}=\frac{\square}{7}$ Subtract the whole numbers. $5-2=\square$ Answer:	
2. Subtract $4\frac{2}{5}-2\frac{3}{5}$. Since $\frac{2}{5}$ is smaller than $\frac{3}{5}$, borrowing is needed. Borrow 1 from the 4. Add 1 as $\frac{5}{5}$ to $\frac{2}{5}$. $\frac{5}{5}+\frac{2}{5}=\frac{\square}{5}$ Subtract the fractions. Subtract the whole numbers. $3\frac{\square}{5}-2\frac{3}{5}=\square\frac{\square}{5}$ Answer:	

Example	Notes
3. Subtract $6\frac{1}{2}-3\frac{1}{4}$. Rewrite the fractions with the LCD. $\frac{1}{2}=\frac{\square}{\square}$, $\frac{1}{4}=\frac{\square}{\square}$ Subtract the fractions. Subtract the whole numbers. Answer:	
4. Subtract $9\frac{1}{3}-2\frac{5}{8}$. Answer:	

Helpful Hints
Borrowing can be needed when subtracting mixed numbers, similar to subtracting integers.

Borrowing one can be rewritten as an improper fraction with the needed denominator, by writing the same number as the numerator.

Concept Check
1. Why must fractions have the same denominator to subtract them?

Practice
Subtract.

2. $2\frac{4}{5}-1\frac{2}{5}$

3. $5\frac{1}{3}-2\frac{2}{3}$

4. $5\frac{1}{3}-2\frac{1}{6}$

5. $6\frac{1}{5}-1\frac{3}{10}$

Name: ______________________________ Date: ________________
Instructor: ___________________________ Section: ______________

Mixed Numbers
4.7 Adding and Subtracting Mixed Numbers - Improper Fractions

Vocabulary

mixed numbers • improper fractions • numerators • lowest common denominators

1. When adding and subtracting mixed numbers, the mixed numbers can be rewritten as ____________.

Step-by-Step Video Notes

Watch the Step-by-Step Video lesson and complete the examples below.

Example	Notes
1. Subtract $3\frac{4}{5} - 2\frac{2}{5}$. Write each mixed number as an improper fraction. $3\frac{4}{5} = \frac{\square}{5}$, $2\frac{2}{5} = \frac{\square}{5}$ Subtract the fractions. $\frac{\square}{5} - \frac{\square}{5} = \frac{\square}{5}$ Change the improper fraction to a mixed number. Answer:	
2. Add $4\frac{2}{3} + 3\frac{2}{3}$. Write each mixed number as an improper fraction and add. $\frac{\square}{3} + \frac{\square}{3} = \frac{\square}{\square}$ Answer:	

Example	Notes
3. Subtract $5\frac{1}{8}-3\frac{3}{8}$. Answer:	
4. Add $1\frac{2}{5}+3\frac{1}{2}$. $\frac{\square}{5}+\frac{\square}{2}$ Rewrite the improper fractions with the LCD. $\frac{\square}{10}+\frac{\square}{10}=\frac{\square}{10}$ Simplify. Answer:	

Helpful Hints

Writing mixed numbers as improper fractions before adding, allows you to avoid changing improper fractions to mixed numbers and carrying a one to the whole number part.

Writing mixed numbers as improper fractions before subtracting, allows you to avoid borrowing from 1 from a whole number and convert it to an improper fraction.

Concept Check

1. What is the first step to adding or subtracting mixed numbers?

Practice

Add.

2. $2\frac{5}{7}+1\frac{3}{7}$

3. $3\frac{1}{4}+1\frac{1}{2}$

Subtract.

4. $4\frac{2}{3}-1\frac{1}{3}$

5. $4\frac{2}{3}-3\frac{1}{6}$

Name: ______________________________ Date: ________________
Instructor: ___________________________ Section: ______________

Operations with Decimals
5.1 Decimal Notation

Vocabulary
place value • standard form • decimal point • decimal fraction

1. A decimal, also known as a decimal number, has three parts: a whole number part, a ____________ ____________, and a decimal part.

Step-by-Step Video Notes
Watch the Step-by-Step Video lesson and complete the examples below.

Example	Notes
1. Write the decimal fraction and the decimal number that represents each shaded part. The decimal fraction for the shaded part is $\frac{\square}{\square}$. The decimal number for the shaded part is $\square$.	
2. Read the decimals. {Insert art showing the place value chart, as on page 3 of video pdf 10.1} 2.4 Read 2.4 as "two and four ____________." 4.17 Read 4.17 as "____________ and ____________ ____________."	

Example	Notes
3. Write each decimal in words. 0.32 0.32 in words is "thirty-two ____________." 15.703 15.703 in words is "____________ and ____________ ____________."	
4. Write each decimal in standard form. twenty-one and two hundred thirty-seven thousandths 21.☐ fourteen and eight hundredths ☐.☐	

Helpful Hints

When reading or writing a decimal number, use the decimal point in place of the word "and."

When writing decimals in standard form, use the given place value to determine the number of decimal places. Write the decimal part in number form so that it ends at the given place value, inserting zeros at the beginning if needed.

Concept Check

1. How many zeros should you insert after the decimal point when writing the number four and three thousandths in standard form?

Practice

Write each decimal in words.

2. 301.03

3. 4.718

Write each decimal in standard form.

4. twenty-seven thousandths

5. five hundred ten and nine tenths

Name: ______________________________ Date: ______________
Instructor: ___________________________ Section: ______________

Operations with Decimals
5.2 Comparing Decimals

Vocabulary
inequality symbols • is less than • is greater than • comparing decimals

1. The symbol < means ____________.

2. ____________ always point to the smaller number.

Step-by-Step Video Notes
Watch the Step-by-Step Video lesson and complete the examples below.

Example	Notes
1. For each pair of decimals, which is larger? 0.14 and 0.41 The whole number part is the same for both. Compare the decimal parts, which have the same place value. $41 > 14$, so ☐ > ☐ 0.6 or 0.59 The decimal parts have different place values. Write all decimals with the same number of decimal places, adding zeros to the ends of decimals as needed. $.60 > .59$, so ☐ > ☐	
2. Fill in the blank with > or < to make a true statement. 4.13 ☐ 4.9 0.983 ☐ 0.7822	

Example	Notes
3. Arrange in order from smallest to largest. 1.23, 1.31, 1.2, 1.0567 The whole number part is the same for all. Write all decimals with the same number of decimal places, adding zeros to the ends of decimals as needed. Compare the digits in each place value, starting from the left. $0 < 2 < 3$, so 1.0567 is the smallest, and ☐ is the largest. 1.2000 is less than 1.2300 Answer:	
4. Order the scores from smallest to largest. During the 2008 Olympic Games, the U.S. Women Gymnastics Team scored 46.875 on the vault, 47.975 on the uneven bars, 47.25 on the balance beam, and 44.425 on the floor exercises. Answer:	

Helpful Hints

Adding zeros to the end of a decimal *does not* change its value. Inserting zeros at the beginning of the decimal part of a number *does* change its value. For example, 0.5 is the same as 0.50, but it is not the same as 0.05.

Concept Check

1. What is a way to compare decimal numbers with a different number of decimal places?

Practice

For each pair of decimals, which is larger?

2. 0.53 or 0.503

3. 6.3 or 6.29

Arrange from smallest to largest.

4. 1.104, 1.04, 1.4, 1.14

5. 0.3, 0.33, 0.303, 0.033

Name: ______________________________ Date: ________________
Instructor: ______________________________ Section: ________________

Operations with Decimals
5.3 Rounding Decimals

Vocabulary
sales tax • batting average • round up • round down

1. You underline the digit in the place value to which you are rounding. If the digit to the right is 5 or more, you will ____________, or increase the underlined digit by 1.

Step-by-Step Video Notes
Watch the Step-by-Step Video lesson and complete the examples below.

Example	Notes
1. Round 13.45812 to the nearest thousandth.	

Whole Numbers				.	Decimals				
One Thousands	Hundreds	Tens	Ones	Decimal Point	Tenths	Hundredths	One – Thousandths	Ten – Thousandths	Hundred – Thousandths

Underline the digit in the place value to which you are rounding.

13.45<u>8</u>12

Look at the digit to the right of the underlined digit. If that digit is 4 or less, round down; if it is 5 or more, round up. That digit is ☐, so round down to 8. Leave off all digits to the right of the underlined digit.

Answer:

Example	Notes
2. Round 6.7481 to the nearest hundredth. Underline the digit in the place value to which you are rounding. 6.7481 Look at the digit to the right of the underlined digit. Round up or round down. Answer:	
3. Round 7.9561 to the nearest tenth. Answer:	

Helpful Hints

Many common statistics, for example grade point average and sports statistics are rounded to a specific place value. Many common money transactions like interest, taxes, and discounts are rounded to the nearest cent.

If the digit you are rounding up is a 9, you change that digit to a 0, and increase the digit to the left by 1. For example, 4.972 rounded to the nearest tenth is 5.0

Leave off all digits to the right of the place to which you are rounding. If you round a decimal to the nearest whole number, do not write a decimal point or any decimal places.

Concept Check

1. Which number has more decimal places, 5.325 rounded to the nearest hundredth, or 6.357895 rounded to the nearest tenth?

Practice

Round to the nearest hundredth.

2. 62.514

3. 99.999

Round to the nearest tenth.

4. 34.96

5. 0.285714

Name: ______________________________ Date: ________________
Instructor: ___________________________ Section: ______________

Operations with Decimals
5.4 Adding and Subtracting Decimals

Vocabulary
estimation • decimal places • decimal points • perimeter of a triangle

1. When adding or subtracting decimals, write the numbers vertically, making sure the ____________ line up.

Step-by-Step Video Notes
Watch the Step-by-Step Video lesson and complete the examples below.

Example	Notes
1. Jennifer spent \$2.97 for a meal and \$1.19 for dessert, including tax. How much did she spend in total? Write the numbers vertically. Make sure the decimal points line up. $\begin{array}{r} 2.97 \\ +1.19 \\ \hline \square.\square\square \end{array}$ Answer:	
2. Add. Check by estimating. $3.27+15.2$ Add zeros as needed so all decimals have the same number of decimal places. $\begin{array}{r} 3.27 \\ +15.2\square \\ \hline \square\square.\square\square \end{array}$ Estimate. Add the whole numbers and estimate the decimal sum. The sum is about 18.5. Answer:	

Example	Notes
3. Subtract. Check by estimating. 21.5 − 16.43 $\begin{array}{r} 21.5\square \\ -16.43 \\ \hline \square\square.\square\square \end{array}$ Answer:	
4. Find the perimeter of the triangle. 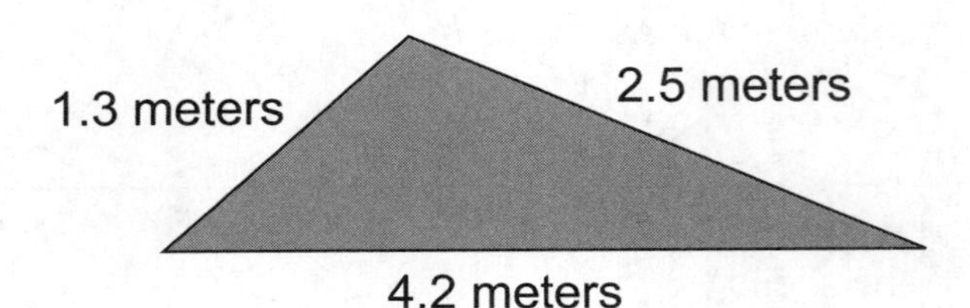 The perimeter of a triangle is the distance around the triangle. To find the perimeter of a triangle, add the lengths of all its sides. $1.3 + 2.5 + 4.2 = \square$ meters	

Helpful Hints
To add and subtract decimals, write the numbers vertically. Line up the decimal points, including the one in the answer. Add zeros to the ends of decimals as needed so all decimals have the same number of decimal places.

Concept Check
1. To subtract $24.3 - 7.52$ vertically, to which number will you insert a zero at the end?

Practice

Add.

2. $5.27 + 12.3$

3. $6 + 11.9 + 4.123$

Subtract.

4. $14.36 - 8$

5. $17.9 - 8.46$

Name: ______________________________ Date: ______________
Instructor: ___________________________ Section: ____________

Operations with Decimals
5.5 Multiplying Decimals

Vocabulary
power of 10 • factor • decimal • whole number

1. To multiply a decimal by a ____________, move the decimal point to the right the same number of places as the number of zeros in the power of 10.

Step-by-Step Video Notes
Watch the Step-by-Step Video lesson and complete the examples below.

Example	Notes
1. Multiply $3.45(6.1)$. Count the total number of decimal places in all factors. Place the decimal point in the answer so that it has this total number of decimal places. $\begin{array}{r} 3.45 \\ \times\ 6.1 \\ \hline 345 \\ \square\square\square\square 0 \\ \hline \square\square\square\square 5 \end{array}$ How many decimal places in all the factors? $\square$ Answer:	
2. Multiply 3.2×0.008. $\begin{array}{r} 3.2 \\ \times\ 0.008 \\ \hline \square.\square\square\square\square \end{array}$ The number of decimal places needed is greater than the number of digits in the answer. Insert zeros before the answer. Answer:	

Example	Notes
3. Multiply 5.793×100. To multiply a decimal by a power of 10, move the decimal point to the right the same number of places as the number of zeros in the power of 10. Move the decimal point 2 places to the right. $5.793 \times 100 =$ 5 7 9 3 Answer:	
4. Ian's new car travels 25.6 miles for each gallon of gas. How many miles can he travel on a full tank, which contains 12 gallons of gas? The answer will have ☐ decimal place(s). 2 5 . 6 × 1 2 Answer:	

Helpful Hints

After multiplying, insert zeros before the answer if the number of decimal places needed is greater than the number of digits in the answer.

When multiplying by powers of 10, if the number of zeros is more than the number of original decimal places, add zero(s) to the end of the decimal in order to move the decimal point the correct number of places.

Concept Check

1. Will you need to add zeros at the end of your answer when multiplying 24.03×100 ?

Practice

Multiply.

2. 3.2×5.08

3. 7.4×0.006

Multiply by the power of 10.

4. 8.136×100

5. 44.12×1000

Name: ______________________ Date: ______________
Instructor: ______________________ Section: ______________

Operations with Decimals
5.6 Dividing Decimals

Vocabulary

divisor • product • dividend • quotient

1. The ____________ is the answer to the division problem.

Step-by-Step Video Notes

Watch the Step-by-Step Video lesson and complete the examples below.

Example	Notes
1. $6.3 \div 3$ Copy the decimal point in the dividend straight up into the quotient. $3\overline{)6.3}$ (decimal point above) Divide. $3\overline{)6.3}$ (box with decimal point above) Answer:	
2. $2.9 \div 1000$ How many zeros are in 1000? ☐ Move the decimal point this same number of spaces to the left; you will need to insert zeros before the whole number in order to move the decimal point the correct number of places. Answer:	

Example	Notes
3. Divide 6.1 by 3. Zeros must be added to the dividend. $\boxed{}$ $3\overline{)6.100}$ Divide until repeating pattern. Answer:	
4. $.0612 \div 0.12$ Move the decimal point to the right until the divisor is a whole number. $0.12 \rightarrow \boxed{}$ Move the decimal place the same number of places in the divisor. $0.612 \rightarrow \boxed{}$ Divide. Answer:	

Helpful Hints
To divide a decimal by a whole number, copy the decimal point straight up into the quotient.

When dividing by a decimal, the decimal point is moved the same number of places in the dividend as needed to achieve a whole number in the divisor.

Concept Check
1. What different steps are taken when dividing a decimal vs. dividing by a decimal?

Practice
Divide.
2. $8.5 \div 5$

3. $5.3 \div 4$

4. $9.4 \div 100$

5. $.0736 \div .23$

Name: ______________________________ Date: ________________
Instructor: ___________________________ Section: ______________

Operations with Decimals
5.7 Order of Operations and Decimals

Vocabulary
addition and multiplication • addition and subtraction • multiplication and division • fraction bars

1. In the order or operations, after evaluating what is inside the parentheses and evaluating any exponents, the next step is to perform _____________ from left to right.

Step-by-Step Video Notes
Watch the Step-by-Step Video lesson and complete the examples below.

Example	**Notes**
1. Simplify $3.6+(0.2+0.3)^2$. Evaluate what is inside the parentheses. $(0.2+0.3)=\square$ Evaluate the exponent. $(\square)^2=\square$ Perform the addition. $3.6+\square=\square$ Answer:	
2. Simplify $(12.6-11.4)^2\div(0.2)(3)$. Evaluate what is inside the parentheses. $\square$ Evaluate the exponent. $\square$ Perform the remaining operations from left to right. Answer:	

Example	Notes
3. Simplify $10.5-3.2+7.2\div 3.6$ Since there are no parentheses or exponents begin by performing the division. $10.5-3.2+\square$ Perform the remaining operations from left to right. Answer:	
4. Simplify $\frac{10.1+2.1(3)}{5.05+3.15}$. Answer:	

Helpful Hints
The order of operations for decimals is the same as for whole numbers.

Treat a fraction bar like there are parentheses around the numerator and the denominator.

Concept Check
1. List the order of operations.

Practice
Simplify.

2. $4.1-(0.5+0.4)^2$

3. $1.3+0.5\div 0.9-0.4$

4. $3+2(0.5)^2$

5. $\frac{0.4+4(.5)}{0.2+0.4}$

Name: ______________________________ Date: ______________
Instructor: ___________________________ Section: ______________

Operations with Decimals
5.8 Converting Fractions to Decimals

Vocabulary
improper fraction • comparing decimals • repeating decimal • equivalent decimal

1. A decimal that repeats in a pattern without end is a(n) ____________.

Step-by-Step Video Notes
Watch the Step-by-Step Video lesson and complete the examples below.

Example	Notes
1. Write $\frac{1}{10}$ as an equivalent decimal. What power of ten is the place value? tenths How many tenths are there? $\square$ Put this answer in the tenths place value after the decimal point. Answer:	
2. Write $\frac{3}{5}$ as a decimal. If the denominator is not a power of 10, then divide the numerator by the denominator. Enter the denominator in the division problem. $\square\overline{)3}$ Perform the division. Answer:	

Example	Notes
3. Convert $\frac{2}{3}$ to a decimal. Set up the long division. $\square\overline{)\square}$ Perform the long division. $\square$ When a decimal repeats in a pattern, place a bar over the repeating number(s). Answer:	
4. Convert $\frac{11}{4}$ to a decimal. Answer:	

Helpful Hints
When converting a fraction to a decimal, divide the numerator by the denominator.

When dividing the numerator by the denominator results in a repeating pattern, placing a bar over the repeating number(s) indicates the repeating decimal.

Concept Check
1. How can a fraction with a denominator that is not power of 10 be converted to a decimal?

Practice
Convert the fraction to a decimal.

2. $\frac{42}{100}$

3. $\frac{5}{8}$

4. $\frac{1}{6}$

5. $\frac{9}{5}$

Name: ______________________________ Date: ______________
Instructor: ____________________________ Section: ______________

Operations with Decimals
5.9 Converting Decimals to Fractions

Vocabulary
improper fraction • denominator • numerator • simplest form

1. Write the decimal part of the decimal number as the ____________ of the fraction.

Step-by-Step Video Notes
Watch the Step-by-Step Video lesson and complete the examples below.

Example	Notes
1. Write 5.277 as a fraction in simplest form. What is the place value of the last 7 in the decimal? ____________ Enter this power of 10 as the denominator in the fraction and write the decimal part of the decimal number as the numerator to the fraction. $\frac{277}{\square}$ Write the fraction or mixed number in simplest form. Answer:	
2. Write 0.96 as a fraction in simplest form. $\frac{96}{\square}$ Simplify by dividing numerator and denominator by common factors. Answer:	

Example	Notes
3. Write 0.0201 as a fraction in simplest form. Answer:	
4. Write 4.0358 as a fraction in simplest form. Answer:	

Helpful Hints
The number of decimal places is equal to the number of zeros in the power of 10 in the denominator. We can also use this to find the denominator.

Always simply the fraction if possible.

Concept Check
1. What are the numerator and denominator when writing a decimal as a fraction?

Practice
Write each decimal as a fraction in simplest form.

2. 0.303
3. 1.75
4. 6.19
5. 0.492

Name: ______________________________ Date: ________________
Instructor: ___________________________ Section: ______________

Ratios, Rates, and Percents
6.1 Ratios

Vocabulary
numerator • denominator • ratio • fraction

1. A(n) _____________ is a comparison of two like quantities, measured in the same units.

Step-by-Step Video Notes
Watch the Step-by-Step Video lesson and complete the examples below.

Example	Notes
1. A gallon of lemonade is made by mixing 2 cups of lemon juice with 14 cups of water. Write the ratio of lemon juice to water in a gallon of lemonade. Use all three forms. The ratio can be written in words. 2 to 14 The ratio can be written using a colon. 2:☐ The ratio can be written as a fraction. $\frac{2}{14}$ Answer:	
2. Write 5 hours to 7 hours as a ratio. Write ratio in word form. ____________ Write the ratio using a colon. ☐ Write the ratio as a fraction. $\frac{\square}{\square}$ Answer:	

Example	Notes
3. Write 4 ounces:32 ounces as a fraction in simplest form. $\frac{4}{\square} = \frac{\square}{\square}$ Answer:	
4. Write \$1.23 to \$3.75 as a fraction in simplest form. Answer:	

Helpful Hints

Ratios can be written in three ways: in words, using a colon, and as a fraction.

We do not need to include the units in ratios since the units are the same and can be divided out as a common factor.

Concept Check

1. How is a ratio written as a fraction?

Practice

Write the ratio in all three forms.

2. An index card measures 3 inches in width and 5 inches in length. Write the ratio of width to length.

3. A classroom has 7 boys and 9 girls. Write the ratio of boys to girls in the class.

Write ratio as a fraction in simplest form.

4. 8 inches to 12 inches

5. \$4.25 to \$6.50

Name: ______________________ Date: ______________
Instructor: ______________________ Section: ______________

Ratios, Rates, and Percents
6.2 Rates

Vocabulary

rate • fraction • ratio • unit rate

1. A(n) ____________ is a ratio that compares two quantities with different units.

2. A(n) ____________ gives the rate for one unit of the item.

Step-by-Step Video Notes

Watch the Step-by-Step Video lesson and complete the examples below.

Example	Notes
1. Write the rate of 240 miles per 10 gallons as a fraction is simplest form. The first quantity is written with units as the numerator. Write the second quantity with units as the denominator. $\dfrac{240 \text{ miles}}{\square \text{ \_\_\_\_\_\_}}$ Simplify. $\dfrac{240 \text{ miles}}{\square \text{ \_\_\_\_\_\_}} = \dfrac{\square \text{ miles}}{\square \text{ gallon}}$ Answer:	
2. Write the rate 18 tomatoes for 4 pots of stew as a fraction in simplest form. Write the ratio as a fraction include units. $\dfrac{18 \text{ \_\_\_\_\_\_\_\_}}{\square \text{ pots of stew}}$ Simplify. Answer:	

Example	Notes
3. Write the ratio 20 door prizes for 110 people as a fraction in simplest form. Remember that units are included in a rate. Answer:	
4. A 6-pack of juice bottles costs $3.00. Find the unit rate of cost per 1 bottle. Write the rate as a fraction with units. ______________ Perform the division and attach the units. Answer:	

Helpful Hints

When writing a rate as a fraction, remember that the first quantity is the numerator and the second quantity is the denominator.

To find a rate, remember to add the units into the fraction.

Concept Check

1. What is the difference between a rate and a unit rate?

Practice

Write the rate as a fraction in simplest form.

2. 320 miles per 8 hours

3. 15 gallons of paint to paint 6 apartments

Find the unit rate.

4. Carl earns $600 for 40 hours of work

5. Kim spent $450 for 15 gallon of paint.

Name: ______________________________ Date: ________________
Instructor: ___________________________ Section: ________________

Ratios, Rates, and Percents
6.3 Proportions

Vocabulary
ratio • equation • proportion • fraction

1. A(n) _____________ is a statement that two quantities are the same, or equal.

2. A(n) _____________ is a statement that two rates or ratios are equal.

Step-by-Step Video Notes
Watch the Step-by-Step Video lesson and complete the examples below.

Example	Notes
1. Write a proportion for 3 is to 6 as 12 is to 24. Write the first ratio as a fraction. $\frac{3}{\square}$ Write the second ratio as a fraction. $\frac{\square}{\square}$ Write an equation representing that the two fractions are equal to each other. Answer:	
2. Determine if $\frac{6}{21} \stackrel{?}{=} \frac{10}{35}$ is a proportion. Find the cross product. $6 \cdot 35 = \square$ Find the cross product. $21 \cdot 10 = \square$ Are these cross products equal? ____________ If equal, then the statement is a proportion. If not, then the statement is not a proportion. Answer:	

Example	Notes
3. Find the missing number in the proportion. $\frac{3}{9} = \frac{4}{?}$ Replace the "?" with an n. $\frac{3}{9} = \frac{4}{n}$ Find the cross products. $3n = \square$ Solve for n by dividing both sides by 3. Answer:	
4. Find the missing number in the proportion. $\frac{5}{6} = \frac{n}{12}$ Answer:	

Helpful Hints

When setting up a proportion, units must "match up" or be in the same place in the fractions.

In a proportion, the cross products must be equal.

Concept Check

1. How do you determine if the statement of a fraction equal to a fraction is a proportion?

Practice

2. Write a proportion for 8 is to 12 as 2 is to 3.

3. Is this statement a proportion?
$\frac{5}{6.2} = \frac{6}{7.3}$

Find the missing number in the proportion.

4. $\frac{6}{18} = \frac{5}{?}$

5. $\frac{7}{8} = \frac{n}{40}$

Name: ______________________________ Date: ________________
Instructor: ____________________________ Section: ______________

Ratios, Rates, and Percents
6.4 Percent Notation

Vocabulary

rate • percent • denominator • fraction

1. ______________ means per 100.

Step-by-Step Video Notes

Watch the Step-by-Step Video lesson and complete the examples below.

Example	Notes
1. Write 14% as a fraction in simplest form. Write the percent number as numerator in a fraction with denominator of 100. $\frac{\square}{100}$ Simplify the fraction. $\frac{\square}{100} = \frac{\square}{\square}$ Answer:	
2. Write $\frac{7}{100}$ as a percent. Since the denominator of the fraction is 100, write the numerator as the percent. $\square$% Answer:	

Example	Notes
3. Is 50% less than, equal to, or greater than one whole unit. 100% represents one whole unit. Answer:	
4. A company produced 2500 personal computers, and 3% of them were defective. How many of the computers were defective? To find 3% of 2500, turn 3% into a fraction. $3\% = \frac{\square}{\square}$ Multiply this fraction by 2500. Answer:	

Helpful Hints

To convert p% to a fraction, write $\frac{p}{100}$.

Remember that 100% represents one whole unit.

Concept Check

1. How is the percent of a number found?

Practice

2. Write 128% as a fraction in simplest form.

3. Write $\frac{29}{100}$ as a percent.

4. Is 225% less than, equal to, or greater than one whole unit?

5. A company has 600 employees, and 52% of them are male. How many employees are male?

Name: ______________________________ Date: ______________
Instructor: ____________________________ Section: ______________

Ratios, Rates, and Percents
6.5 Percent and Decimal Conversions

Vocabulary

decimal • percent • ratio • fraction

1. ____________ means per 100.

Step-by-Step Video Notes

Watch the Step-by-Step Video lesson and complete the examples below.

Example	Notes
1. Write 51% as a decimal. Place the decimal at the end of the number and then move it two places to the left. [.] The percent sign is left off. Answer:	
2. Write 4.5% as a decimal. Answer:	
3. Write 0.71 as a percent. Move the decimal two places to the right, adding zeros if needed. 0.71 → [ ] Write the percent sign. Answer:	

Example	Notes
4. Write 2.3 as a percent. Answer:	

Helpful Hints
When converting a percent to a decimal, it may be necessary to add zeros.

When converting a decimal to a percent, it may be necessary to add zeros.

Concept Check
1. What are the similarities and differences in converting a percent to a decimal and converting a decimal to a percent?

Practice

Write the % as a decimal.

2. 32.5%

3. 6.1%

Write the decimal as a percent.

4. 1.8

5. 0.025

Name: ______________________ Date: ______________

Instructor: ______________________ Section: ______________

Ratios, Rates, and Percents
6.6 Percent and Fraction Conversions

Vocabulary

decimal • percent • ratio • fraction

1. If the denominator of a fraction is 100, then the numerator is the ____________.

Step-by-Step Video Notes

Watch the Step-by-Step Video lesson and complete the examples below.

Example	Notes
1. Write 8% as a fraction in simplest form. Write the percent as the numerator with the denominator as 100. $\frac{\square}{100}$ Simplify. Answer:	
2. The sales tax is 4.5% of the price. Write the percent as a fraction in simplest form. Write the percent as an equivalent fraction. $4.5 = \frac{\square}{10}$ Multiply this fraction by $\frac{1}{100}$. $\frac{\square}{10} \cdot \frac{1}{100} = \frac{\square}{\square}$ Answer:	

Example	Notes
3. Write $\frac{5}{8}$ as a percent. Divide 5 by 8. Convert to a percent by moving the decimal place two places to the right and then write the percent sign. Answer:	
4. Write $\frac{4}{7}$ as a percent. Round to the nearest tenth of a percent. Answer:	

Helpful Hints

You can convert p% to a fraction by dividing p by 100 or by multiplying p by $\frac{1}{100}$.

Remember that when a fraction has a denominator of 100, the numerator is the percent.

Concept Check

1. What are the similarities and differences in converting a percent to a fraction and converting a fraction to a percent?

Practice

Write the % as a fraction in simplest form.

2. 12%

3. 6.1%

Write the fraction as a decimal. Round to the nearest tenth of a percent as needed.

4. $\frac{3}{5}$

5. $\frac{5}{6}$

Name: ______________________________ Date: ______________
Instructor: ____________________________ Section: ______________

Ratios, Rates, and Percents
6.7 The Percent Equation

Vocabulary

decimal • percent equation • base • fraction

1. The ____________ states "the amount is equal to the percent times the base."

Step-by-Step Video Notes

Watch the Step-by-Step Video lesson and complete the examples below.

Example	Notes
1. What amount is 30% of 140? Substitute the values into the percent equation, amount = $\% \times$ base Enter 30% as a decimal. $a = \square \times 140$ Multiply to find a. Answer:	
2. 34 is 50% of what number? Substitute the values into the percent equation, amount = $\% \times$ base Enter 50% as a decimal. $34 = \square \times b$ Solve the equation for b by dividing each side of the equation by the decimal in front of b. Answer:	

Example	Notes
3. 45 is what percent of 180? Substitute the values into the percent equation, $\text{amount} = \% \times \text{base}$, letting p stand for the unknown percent. $\square = p \times \square$ Solve for p. Answer:	
4. 10 is what percent of 30? Round to the nearest tenth of a percent. Answer:	

Helpful Hints
Usually the base appears after the word "of."

In math "is" translates to an equal sign, and "of" translates into multiplication.

Concept Check

1. State the percent equation and how it can be used to find each of one of the three parts, provided the other two parts are given.

Practice
Use the percent equation to find the following. Round to the nearest tenth of a percent as needed.

2. What amount is 40% of 150?
3. 24 is 20% of what number?
4. 32 is what % of 160?
5. 25 is what percent of 86?

Name: ______________________________ Date: ______________

Instructor: ____________________________ Section: ______________

Ratios, Rates, and Percents
6.8 The Percent Proportion

Vocabulary

amount • percent • ratio • fraction

1. The ______________ states "the ratio of the amount to the base is equal to the ratio of the percent $\frac{p}{100}$."

Step-by-Step Video Notes

Watch the Step-by-Step Video lesson and complete the examples below.

Example	Notes
1. What amount is 20% of 85? Substitute the values into the percent proportion, $\frac{\text{amount}}{\text{base}} = \frac{p}{100}$. $\frac{a}{\square} = \frac{20}{100}$ Cross multiply. $100(a) = \square$ Solve for a by dividing each side of the equation by 100. Answer:	
2. 21 is 10.5% of what number? Substitute the values into the percent proportion, $\frac{\text{amount}}{\text{base}} = \frac{p}{100}$. $\frac{\square}{b} = \frac{\square}{100}$ Cross multiply. $(\square)b = (\square)(100)$ Solve for b. Answer:	

Example	Notes
3. 83 is what percent of 332? Substitute the values into the percent proportion, $\frac{\text{amount}}{\text{base}} = \frac{p}{100}$. Solve for p. Answer:	
4. 4 is what percent of 48? Round to the nearest tenth of a percent. Answer:	

Helpful Hints

The percent proportion is an alternative to the percent equation.

Usually the base appears after the word "of."

Concept Check

1. State the percent proportion and how it can be used to find each of one of the three parts, provided the other two parts are given.

Practice

Use the percent proportion to find the following.

Round to the nearest tenth as necessary.

2. What amount is 30% of 160?

3. 33 is 11% of what number?

4. 96 is what % of 384?

5. 6 is what percent of 96?

Name: ______________________________ Date: ______________
Instructor: ____________________________ Section: ______________

Ratios, Rates, and Percents
6.9 Percent Applications

Vocabulary
amount • percent • base • fraction

1. The percent equation states "the amount is equal to the percent times the ____________."

2. The percent proportion states "the ratio of the amount to the ____________ is equal to the ratio of the percent $\frac{p}{100}$."

Step-by-Step Video Notes
Watch the Step-by-Step Video lesson and complete the examples below.

Example	Notes
1. Eli purchased a wrist watch for \$60. If the sales tax rate is 7%, how much sales tax did Eli pay? What was the total cost of the watch? Substitute the values into the application percent equation: $\begin{pmatrix}\text{amount of}\\ \text{sales tax}\end{pmatrix}=\begin{pmatrix}\text{tax rate}\\ \text{as a decimal}\end{pmatrix}\times\begin{pmatrix}\text{cost of}\\ \text{item}\end{pmatrix}$ $a=(.07)\times\square$ Multiply to solve for a. $a=\square$ Substitute values and perform addition to find total cost. $\begin{pmatrix}\text{total cost}\\ \text{of item}\end{pmatrix}=\begin{pmatrix}\text{amount of}\\ \text{sales tax}\end{pmatrix}+\begin{pmatrix}\text{cost of}\\ \text{item}\end{pmatrix}$ Answers:	

Example	Notes
2. Lucky's Shoes is having a 30% discount shoe sale. How much will Jan save on a $40 pair of shoes? What is the sales price of the shoes? Use the percent discount equation: $\begin{pmatrix}\text{amount of}\\ \text{discount}\end{pmatrix} = \begin{pmatrix}\text{\% discount}\\ \text{as a decimal}\end{pmatrix} \times \begin{pmatrix}\text{cost}\\ \text{of item}\end{pmatrix}$ Use amount of discount to find sales price. $\begin{pmatrix}\text{sales}\\ \text{price}\end{pmatrix} = \begin{pmatrix}\text{original}\\ \text{price}\end{pmatrix} - \begin{pmatrix}\text{amount of}\\ \text{discount}\end{pmatrix}$ Answer:	

Helpful Hints
If you know the percent and the base when solving a percent application problem, you should use the percent equation.

Remember to use the percent as a decimal in the percent equation.

Concept Check
1. How does finding the amount of sales tax of an item make use of the percent equation?

Practice
Samantha wants to buy a $40 sweater. Find the following.

2. What is the amount of sales tax if the sales tax rate is 5%?

3. What is the total cost of the sweater if the sales tax rate is 5%

4. What is the amount of discount if there is a 15% discount sale?

5. What is the sale price if Samantha buys the sweater during the 15% discount sale?

Name: ______________________________ Date: ______________
Instructor: ____________________________ Section: ______________

U.S. and Metric Measurement
7.1 U.S. Length

Vocabulary

conversions • unit fraction • length • mixed units

1. A ____________ has a value of 1. Its numerator and denominator express the same value in different ways.

Step-by-Step Video Notes

Watch the Step-by-Step Video lesson and complete the examples below.

Example	Notes
1. Convert 24 yards to feet using unit fractions.	

Table 2 U.S. Length Conversions
12 in. = 1ft
36 in. = 1 yd
3 ft = 1 yd
5280 ft = 1 mi
1760 yd = 1 mi

There are ☐ ft in 1 yd. Using $\frac{\text{unit of measurement converting to}}{\text{original unit of measurement}}$, the numerator is in ft, and the denominator is in yd.

$$24 \text{ yards} \cdot \frac{\square \text{ feet}}{1 \text{ yard}} = \square \text{ feet}$$

2. Convert 15,840 ft to mi using unit fractions.

There are ☐ feet in 1 mile.

$$15{,}840 \text{ ft} \cdot \frac{1 \text{ mi}}{\square \text{ ft}} = \square \text{ mi}$$

Example	Notes
3. Convert the following using unit fractions. 90 in. to yd 90 in. $\cdot \dfrac{\square\ \_\_\_\_\_\_\_}{\square\ \_\_\_\_\_\_\_} = \square$ yd Answer:	
4. Gage is 54 inches tall. Convert this to feet and inches. 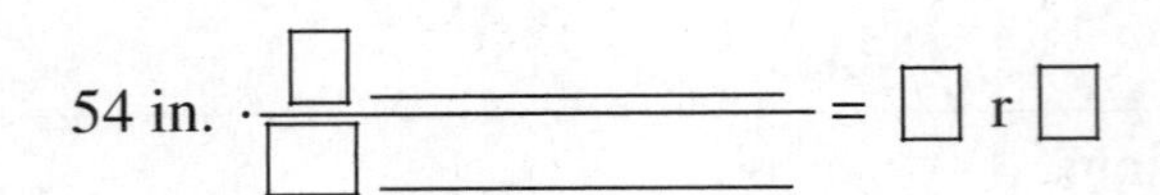 The quotient is the number of feet, and the remainder is in inches. Answer:	

Helpful Hints

Use unit fractions such as $\dfrac{12\text{ in.}}{1\text{ ft}}$, $\dfrac{36\text{ in.}}{1\text{ yd}}$, $\dfrac{3\text{ ft}}{1\text{ yd}}$, $\dfrac{5280\text{ ft}}{1\text{ mi}}$, and $\dfrac{1760\text{ yd}}{1\text{ mi}}$ to convert units.

The fraction $\dfrac{\text{unit of measurement converting to}}{\text{original unit of measurement}}$ can be helpful to use when converting units.

Concept Check

1. Why is a unit fraction equal to 1 if the numerator and denominator have different numbers?

Practice

Convert the following using unit fractions.

2. 468 in. to yd

3. 3.1 miles to yards

Convert the heights to feet and inches.

4. Ben is 68 inches tall.

5. Gomez is 76 inches tall.

Name: ______________________________ Date: ________________
Instructor: ___________________________ Section: ______________

U.S. and Metric Measurement
7.2 U.S. Weight and Capacity

Vocabulary

weight • mass • capacity • pound • gallon

1. ____________ is related to the gravitational pull on an object.

2. ____________ is the amount of space inside a three-dimensional figure.

Step-by-Step Video Notes

Watch the Step-by-Step Video lesson and complete the examples below.

Example	Notes
1. Convert 7.5 tons to pounds. **U.S. Weight Conversions** 16 oz = 1 lb 2000 lbs = 1 ton There are ☐ pounds in 1 ton. Using $\frac{\text{unit of measurement converting to}}{\text{original unit of measurement}}$, the numerator is in pounds, and the denominator is in tons. $7.5 \text{ tons} \cdot \frac{\square \text{ pounds}}{1 \text{ ton}} = \square \text{ pounds}$ Answer:	
2. Convert 64 oz to lb. There are ☐ oz in 1 lb. $64 \text{ oz} \cdot \frac{1 \text{ \_\_\_\_}}{\square \text{ oz}} = \square\ \square$ Answer:	

Example	Notes
3. Convert 26 quarts to gallons.	

Table 4 U.S. Capacity Conversions
8 fluid ounces = 1 cup
2 cups = 1 pint
16 fluid ounces = 1 pint
2 pints = 1 quart
4 quarts = 1 gallon

Give your answer in decimal or fraction form.

Answer:

4. Lashonda buys a bottle of ketchup that contains 44 fl oz. Convert this to pints and fluid ounces.

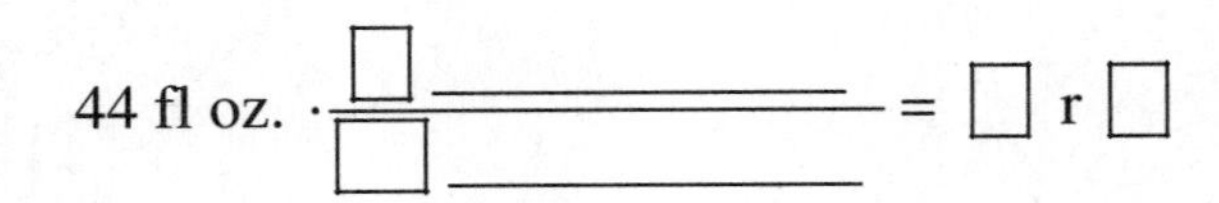

The quotient is the number of pints, and the remainder is in fluid ounces.

Answer:

Helpful Hints

Use unit fractions such as $\frac{16 \text{ oz}}{1 \text{ lb}}$, $\frac{2000 \text{ lb}}{1 \text{ ton}}$, $\frac{8 \text{ fl oz}}{1 \text{ cup}}$, $\frac{16 \text{ fl oz}}{1 \text{ pint}}$, $\frac{1 \text{ quart}}{2 \text{ pints}}$, and $\frac{4 \text{ quarts}}{1 \text{ gallon}}$ to convert units. Note that weight and capacity are often given in mixed units, such as pounds and ounces, pints and fluid ounces, etc.

Concept Check

1. Why use pounds and ounces for the weight of a newborn baby, rather than a decimal number, as is often used with weights measured in tons?

Practice

Convert the following to pounds.

2. 96 ounces

3. 3.7 tons

Convert the following to gallons.

4. 64 fl oz

5. 36 quarts

Name: ______________________________ Date: ______________
Instructor: ____________________________ Section: ______________

U.S. and Metric Measurement
7.3 Metric Length

Vocabulary

metric prefixes • meter • milli- • centi- • kilo-
deka- • hecto- • deci-

1. The metric prefix ____________ means 1000.

Step-by-Step Video Notes

Watch the Step-by-Step Video lesson and complete the examples below.

Example	Notes

1. Convert 400 centimeters to meters.

Table 1 Metric Units of Length

Conversion	Unit Fraction
1000 millimeters (mm) = 1 meter (m)	$\frac{1 \text{ meter}}{1000 \text{ millimeters}}$ or $\frac{1000 \text{ millimeters}}{1 \text{ meter}}$
100 centimeters (cm) = 1 meter (m)	$\frac{1 \text{ meter}}{100 \text{ centimeters}}$ or $\frac{100 \text{ centimeters}}{1 \text{ meter}}$
10 decimeters (dm) = 1 meter (m)	$\frac{1 \text{ meter}}{10 \text{ decimeters}}$ or $\frac{10 \text{ decimeters}}{1 \text{ meter}}$
1 meter (m) is the basic unit of length	
10 meters (m) = 1 dekameter (dam)	$\frac{10 \text{ meter}}{1 \text{ dekameters}}$ or $\frac{1 \text{ dekameters}}{10 \text{ meter}}$
100 meters (m) = 1 hectometer (hm)	$\frac{100 \text{ meter}}{1 \text{ hectometer}}$ or $\frac{1 \text{ hectometer}}{100 \text{ meter}}$
1000 meters (m) = 1 kilometer (km)	$\frac{1000 \text{ meter}}{1 \text{ kilometer}}$ or $\frac{1 \text{ kilometer}}{1000 \text{ meter}}$

There are ☐ centimeters in 1 meter.

$$400 \text{ centimeters} \cdot \frac{1 \text{ meter}}{100 \text{ centimeters}} = \square \text{ meters}$$

Answer:

Example	Notes
2. Convert 96.3 km to m. List the prefixes. km hm dam m dm cm mm To convert from km to m, move the decimal point ☐ places to the ____________. Answer:	
3. Convert 150 dam to cm. 150 dam = ☐☐☐,☐☐☐ cm Answer:	

Helpful Hints

The metric system is based on powers of 10. Converting units can be done by moving the decimal point. The mnemonic "Kangaroos hopping down mountains drinking chocolate milk" can help you remember the metric prefixes in order from largest to smallest, kilometers, hectometers, decameters, meter, decimeter, centimeter, millimeter.

List the prefixes like this km hm dam m dm cm mm is a visual way to tell which way to move the decimal point when converting metric units of length. Start with the original unit and move to the new unit. Move the decimal point accordingly, the same number of spaces and in the same direction, adding zeros as necessary.

Concept Check

1. When converting from meters to centimeters, how many places and in what direction should you move the decimal point?

Practice

Convert the following to meters.

2. 87 cm

3. 6.2 km

Convert the following to centimeters.

4. 25 m

5. 44 mm

Name: ______________________________ Date: ______________

Instructor: ______________________________ Section: ______________

U.S. and Metric Measurement
7.4 Metric Mass and Capacity

Vocabulary

gram • mass • kilogram • milligram • dekagram
liter • milliliter • capacity • deciliter • kiloliter

1. The basic unit of mass in the metric system is the ____________.

2. A ____________ is slightly more than two pounds.

Step-by-Step Video Notes

Watch the Step-by-Step Video lesson and complete the examples below.

Example	Notes
1. Convert 32 centigrams to grams. There are ☐ centigrams in 1 gram. Using $\dfrac{\text{unit of measurement converting to}}{\text{original unit of measurement}}$, the numerator is in grams, and the denominator is in centigrams. $32 \text{ cg} \cdot \dfrac{1 \text{ g}}{\square \text{ cg}} = \square \text{ g}$ Answer:	
2. Convert 216 kg to cg. List the prefixes. kg hg dag g dg cg mg To convert from kg to cg, move the decimal point ☐ places to the ____________. Add zero(s) to the end of the decimal to move the decimal point the correct number of places. Answer:	

Example	Notes
3. Convert 900 mL to L. List the prefixes. kL hL daL L dL cL mL To convert from mL to L, move the decimal point ☐ places to the ____________. Answer:	
4. Convert 83.2 L to cL. Answer:	

Helpful Hints

Regardless of the type of measure (length, mass, or capacity), the metric prefixes always have the same meaning and relationship to the basic unit.

With mass and capacity, the prefixes kilo- and milli- are most often used.

Concept Check

1. Which metric unit is closest to a quart in U.S. measurement? Is a gallon more or less than 4 liters?

Practice

Convert the following to grams.

2. 96 mg

3. 5.8 kilograms

Convert the following to liters.

4. 500 mL

5. 4.4 kL

Name: ______________________________ Date: ______________
Instructor: ______________________________ Section: ______________

U.S. and Metric Measurement
7.5 Converting between U.S. and Metric Units

Vocabulary
meter • gallon • pound • approximately

1. The symbol ≈ means ____________.

Step-by-Step Video Notes
Watch the Step-by-Step Video lesson and complete the examples below.

Example	Notes
1. Convert 44.02 miles to kilometers.	

Table 1 U.S. to Metric (Length)

Conversion	Unit Fraction
1 inch (in.) = 2.54 centimeters (cm)	$\frac{2.54 \text{ cm}}{1 \text{ in.}}$
1 foot (ft) = 0.30 meter (m)	$\frac{0.30 \text{ m}}{1 \text{ ft}}$
1 yard (yd) = 0.91 meter (m)	$\frac{0.91 \text{ m}}{1 \text{ yd}}$
1 mile (mi) = 1.61 kilometers (km)	$\frac{1.61 \text{ km}}{1 \text{ mi}}$

Table 2 Metric to U.S. (Length)

Conversion	Unit Fraction
1 meter (m) = 39.37 inches (in.)	$\frac{39.37 \text{ in.}}{1 \text{ m}}$
1 meter (m) = 1.09 yards (yd)	$\frac{1.09 \text{ yd}}{1 \text{ m}}$
1 meter (m) = 3.28 feet (ft)	$\frac{3.28 \text{ ft}}{1 \text{ m}}$
1 kilometer (km) = .62 mile (mi)	$\frac{62 \text{ mi}}{1 \text{ km}}$

There are about ☐ kilometers in 1 mile.

Convert using the unit fraction.

$$44.02 \text{ mi} \cdot \frac{1.61 \text{ km}}{1 \text{ mi}} = \square \text{ km}$$

Answer:

Example	Notes
2. Convert 5.2 kg to lb. Use the unit fraction	

Table 3 U.S. to Metric (Weight/Mass)

Conversion	Unit Fraction
1 ounce (oz) = 28.35 grams (g)	$\frac{28.35 \text{ g}}{1 \text{ oz}}$
1 pound (lb) = 0.45 kilogram (kg)	$\frac{0.45 \text{ kg}}{1 \text{ lb}}$

Table 4 Metric to U.S. (Weight/Mass)

Conversion	Unit Fraction
1 kilogram (kg) = 2.20 pounds (lb)	$\frac{2.20 \text{ lb}}{1 \text{ kg}}$
1 gram (g) = 0.035 ounce (oz)	$\frac{0.035 \text{ oz}}{1 \text{ g}}$

Answer:

3. Convert 3 qt to L. Use the unit fraction to convert. Round to the nearest tenth.

Table 5 U.S. to Metric (Capacity)

Conversion	Unit Fraction
1 quart (qt) = 0.95 liter (L)	$\frac{0.95 \text{ L}}{1 \text{ qt}}$
1 gallon (gal) = 3.79 liters (L)	$\frac{3.79 \text{ L}}{1 \text{ gal}}$

Table 6 Metric to U.S. (Capacity)

Conversion	Unit Fraction
1 liter (L) = 1.06 quarts (qt)	$\frac{1.06 \text{ qt}}{1 \text{ L}}$
1 liter (L) = 0.26 gallon (gal)	$\frac{0.26 \text{ gal}}{1 \text{ L}}$

Answer:

Helpful Hints

Almost all unit fractions used to convert between U.S. and metric units are approximate. Not all tables provide every conversion fact. You may need to change to units provided in the table, then use another conversion fact you know to complete the conversion.

Concept Check

1. When measuring a length, will there be more units if you measure in yards or in meters?

Practice

Convert the following to meters.

2. 100 yards
3. 6.2 miles

Convert the following to quarts.

4. 18 L
5. 947 mL

Name: ______________________________ Date: ______________
Instructor: ___________________________ Section: ______________

U.S. and Metric Measurement
7.6 Time and Temperature

Vocabulary
conversion • unit fraction • denominator • equivalent fraction

1. A ____________ has a value of 1. Its numerator and denominator express the same value in different ways.

Step-by-Step Video Notes
Watch the Step-by-Step Video lesson and complete the examples below.

Example	Notes
1. Convert 195 minutes to hours. Use the unit fraction, where the numerator is hours and the denominator is minutes: $\dfrac{\text{unit of measurement converting to}}{\text{original unit of measurement}}$ Fill in with the appropriate values. $\dfrac{1 \text{ hour}}{\square \text{ minutes}}$ Multiply 195 minutes by the unit fraction. Answer:	
2. Convert 4.4 hours to seconds. First convert hours to minutes using the unit fraction. $4.4 \text{ hours} \times \dfrac{\square \text{ minutes}}{1 \text{ hour}} = \square \text{ minutes}$ Next convert these minutes to seconds. Answer:	

Example	Notes
3. Convert 36° C to degrees Fahrenheit. Use the following formula: $F = 1.8 \times C + 32$ Fill in degrees Celsius. $F = 1.8(\square) + 32$ Solve for F by following order of operations. Answer:	
4. Convert 81° F to degrees Celsius. Use the following formula: $C = \dfrac{5 \times F - 160}{9}$ Answer:	

Helpful Hints

The fraction $\dfrac{\text{unit of measurement converting to}}{\text{original unit of measurement}}$ is helpful when converting units of time.

When converting units of time, sometimes more than one unit fraction will be needed.

Concept Check

1. Why is a unit fraction equal to 1 if the numerator and denominator have different numbers?

Practice

Convert the following measurements of time using unit fractions.

2. 145 minutes to hours

3. 1.5 hours to seconds

Convert the following measurements of temperature using the appropriate formula.

4. 100° C to degrees Fahrenheit

5. 72°F to degrees Celsius

Name: ______________________________ Date: ______________

Instructor: ____________________________ Section: ______________

Introduction to Geometry
8.1 Lines and Angles

Vocabulary

point • line • line segment • ray • angle • measurement • right angle
acute angle • straight angle • obtuse angle • complementary • supplementary

1. A(n) _____________ is a portion of a line with two endpoints.

2. A(n) _____________ is a portion of a line with one endpoint.

3. A(n) _____________ is an angle that measures 90°.

4. Two angles are _____________ if the sum of their measures is 180°.

Step-by-Step Video Notes

Watch the Step-by-Step Video lesson and complete the examples below.

Example	Notes
1. Identify the following figure as a line, line segment, or a ray and give the name. A B Use the endpoint(s) and the appropriate symbol to name the figure. Answer:	
2. Identify the angle shown below as a straight angle, right angle or neither. 180° A B C Answer:	

Example	Notes
3. Identify the given angle as acute, obtuse, or neither. (figure: angle ABC with points A, B, C) Review the definitions of acute angle and obtuse angle. Answer:	
4. Find the complement of a 65° angle. Two angles are complementary if the sum of their angles is [°]. Subtract 65° from the sum. Answer:	

Helpful Hints

If two complementary angles are adjacent, they will form a right angle.

If two supplementary angles are adjacent, they will form a straight angle.

Concept Check

1. Describe how 90° and 180° are used to define the following angles: acute, right, obtuse, straight, complementary and supplementary.

Practice

Identify the type of angle described as acute, right, or obtuse.

2. An angle measuring 65°

3. An angle measuring 90°

Find the following angles.

4. The supplement of 120°

5. The complement of 35°

Name: ______________________________ Date: ________________
Instructor: ___________________________ Section: ______________

Introduction to Geometry
8.2 Figures

Vocabulary

triangle • right triangle • acute triangle • obtuse triangle
polygon • quadrilateral • parallel lines • rectangle
square • trapezoid • parallelogram • parallel lines

1. A(n) _____________ is a four-sided geometric figure.

2. A(n) _____________ is a quadrilateral with opposite sides that are equal in length and four angles that are right angles.

3. A(n) _____________ is a quadrilateral with only one pair of opposite sides that are parallel.

Step-by-Step Video Notes
Watch the Step-by-Step Video lesson and complete the examples below.

Example	Notes
1. If two angles of a triangle measure 55° and 70°, find the measure of the third angle. Then identify the triangle as acute, right, or obtuse. Find the sum of the two given angles. $\square$° Subtract this sum from 180. $180 - \square = \square$ Is there an obtuse angle? _____________ Is there a right angle? _____________ Are all the angles acute? _____________ Answer:	

Example	Notes
2. Identify the figure below. Answer:	
3. Identify the figure below. Answer:	

Helpful Hints

The sum of the angles of any triangle is 180°.

Every square is also a rectangle, but every rectangle is not a square.

Concept Check

1. What are the similarities and differences among these quadrilaterals: rectangle, square, trapezoid, parallelogram, diamond and kite.

Practice

2. Find the third angle of a triangle having angles of 100° and 50°.

3. Identify the figure below.

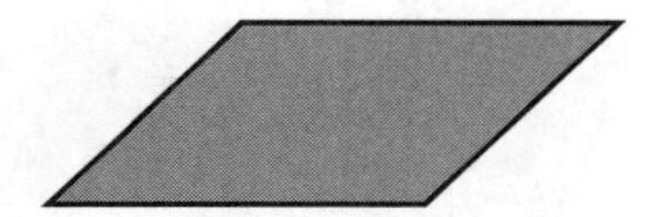

4. Identify a triangle with angles of 95°, 45°, and 40° as acute, right or obtuse.

5. Identify the figure below.

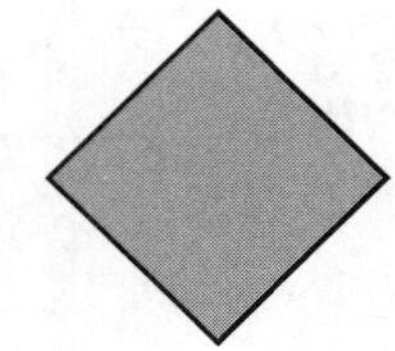

Name: ______________________________ Date: ______________
Instructor: ____________________________ Section: ______________

Introduction to Geometry
8.3 Perimeter – Definitions and Units

Vocabulary
polygon • distance • perimeter • parallelogram

1. The ____________ of a figure can be found by adding the lengths of all its sides.

Step-by-Step Video Notes
Watch the Step-by-Step Video lesson and complete the examples below.

Example	Notes
1. Which, if any, of these situations, painting a room, spreading grass seed, or hanging gutters around a house would involve perimeter? Is painting a room, the distance around the room? ____________ Is spreading grass seed, only around the edge of the yard? ____________ Is hanging gutters around the house, around the edge of the house? ____________ Answer:	
2. Find the perimeter of the figure. 6 inches 8 inches 4 inches Add the lengths of the three sides of the figure. 6 inches + ☐ inches + ☐ inches = ☐ inches Answer:	

Example	Notes
3. Find the perimeter of the figure.	

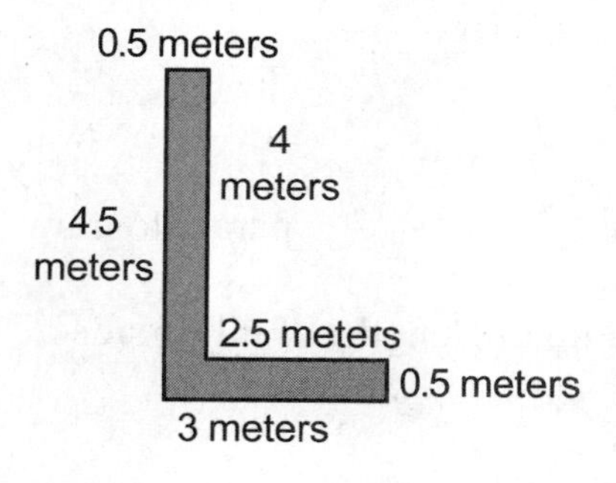

Answer:

Helpful Hints

Perimeter is always measured in units of length.

Concept Check

1. What is the similarity in finding the perimeter of a rectangle and a triangle?

Practice

2. Which of these units, if any, could represent perimeter? Square feet, miles, or cubic meters

3. Find the perimeter of the figure.

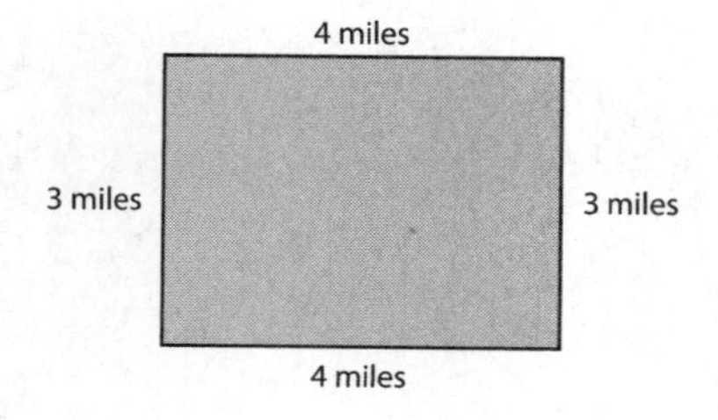

4. Find the perimeter of the figure.

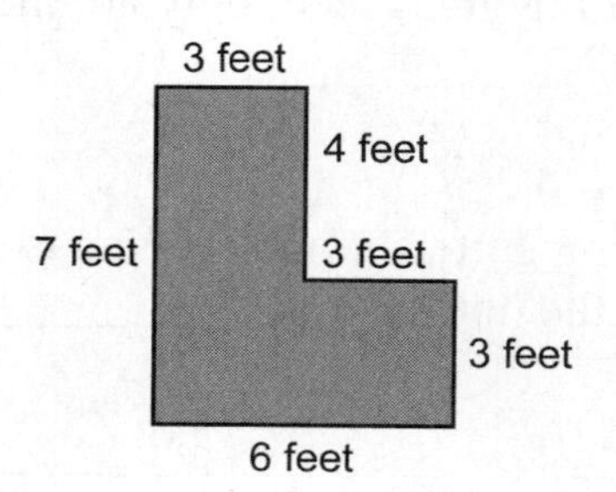

5. Find the perimeter of the figure.

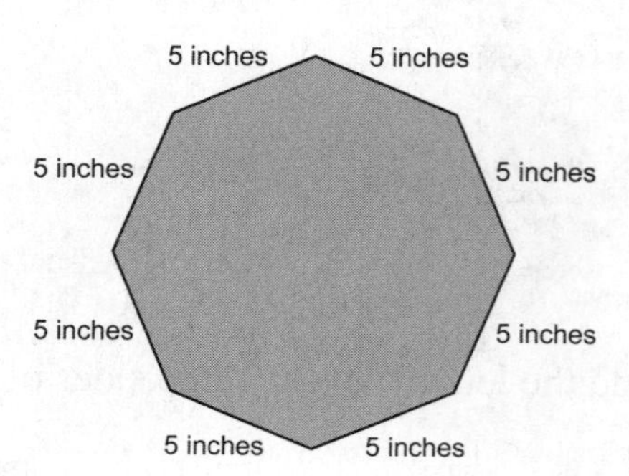

Name: ______________________________ Date: ________________
Instructor: ___________________________ Section: ______________

Introduction to Geometry
8.4 Finding Perimeter

Vocabulary
angle • square • perimeter • rectangle

1. The ____________ of a figure can be found by adding the lengths of all its sides.

Step-by-Step Video Notes
Watch the Step-by-Step Video lesson and complete the examples below.

Example	Notes
1. Find the perimeter of the triangle below. 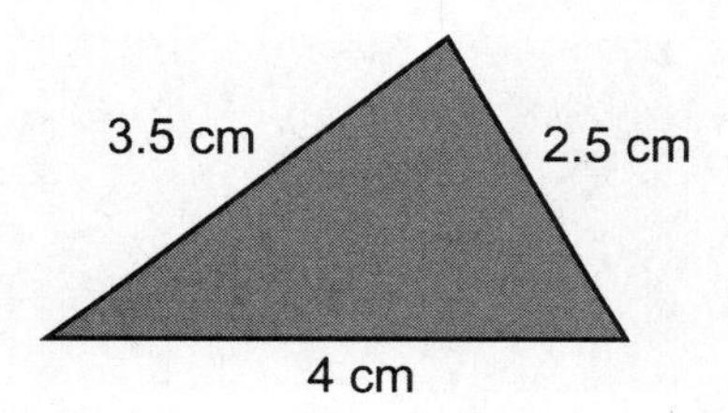 Use the perimeter formula: $P = a + b + c$, where a, b, and c represent the sides of the triangle. $P = 3.5 \text{ cm} + \square \text{ cm} + \square \text{ cm}$ Find the sum. Answer:	
2. Find the perimeter of the rectangle below.  Use the perimeter formula: $P = 2l + 2w$, where l represents length and w represents width. Answer:	

Example	Notes
3. Find the perimeter of the figure. 14 m 14 m Use the formula: $P = 4s$, where s represents the side length of the square. Answer:	
4. Find the perimeter of a triangular garden with side lengths of 2 feet, 5 feet, and 10 feet. Answer:	

Helpful Hints
Perimeter is always measured in units of length.

The perimeter can be found by adding the lengths of the sides of a figure, but formulas for a triangle, a square and a rectangle can be used.

Concept Check
1. What are the perimeter formulas for a triangle, a rectangle, and a square?

Practice
Use the appropriate perimeter formula to find the perimeter of the figures shown or described below.

2.

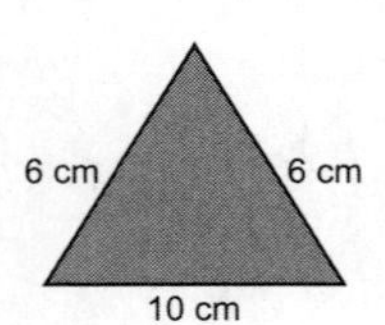

3. A rectangular desk with length 4 feet and width 2.5 feet.

4.

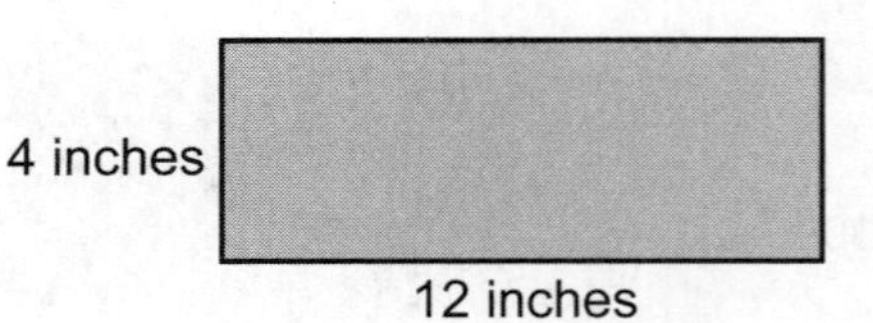

5. A square bandana with side lengths of 8 inches.

Name: ______________________________ Date: ________________
Instructor: ___________________________ Section: ______________

Introduction to Geometry
8.5 Area – Definitions and Units

Vocabulary
square unit • area • perimeter • square inches

1. The ____________ of a figure is the measure of the surface inside the figure, which is measured in square units.

Step-by-Step Video Notes
Watch the Step-by-Step Video lesson and complete the examples below.

Example	Notes
1. How many square units are needed to cover the figure below completely? 4 inches; 3 inches; 3 inches; 4 inches The small squares measuring 1 in.×1 in. are square units. Count the number of square units in this figure. Answer:	
2. Draw two different rectangles each of which has an area of 6 square units.	

Example	Notes
3. Find the area of a right triangle with a base and height of 4 in.	

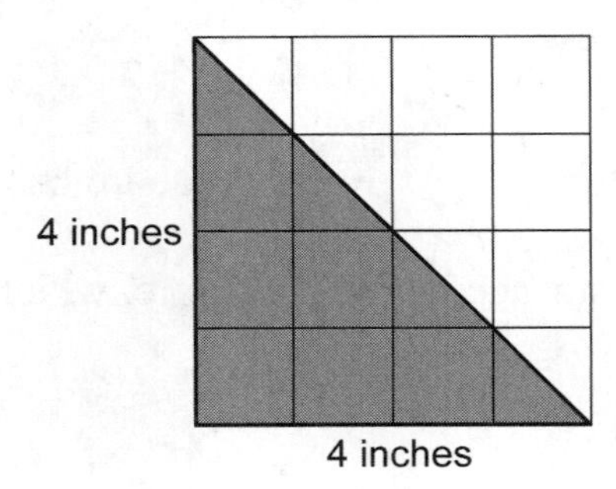

Add the half squares to make whole squares.

Answer:

Helpful Hints

Area is always measured in square units of length.

Two figures can have different shapes, but have the same area.

Concept Check

1. How is a grid of square units used to determine the area of a figure?

Practice

2. Find the area of a right triangle with a base and height of 3 in.

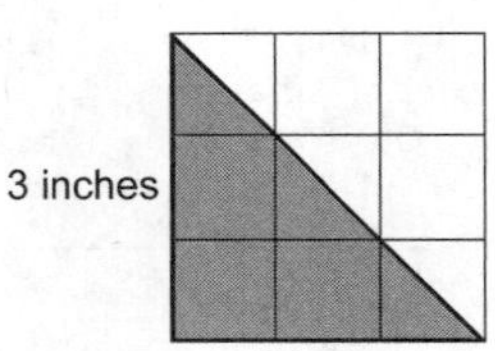

3. How many square units are needed to cover the figure below completely?

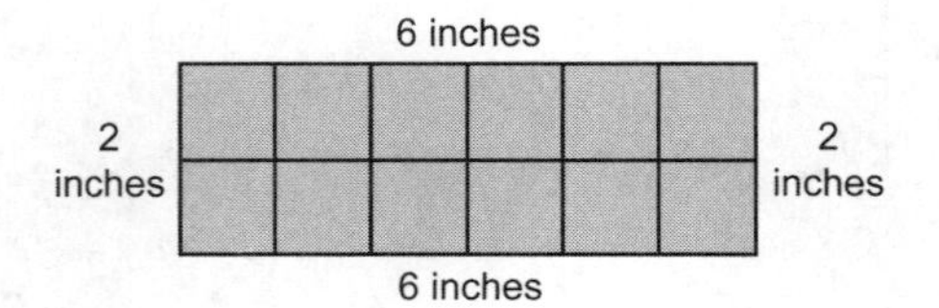

4. Draw a rectangle which has an area of 12 square units.

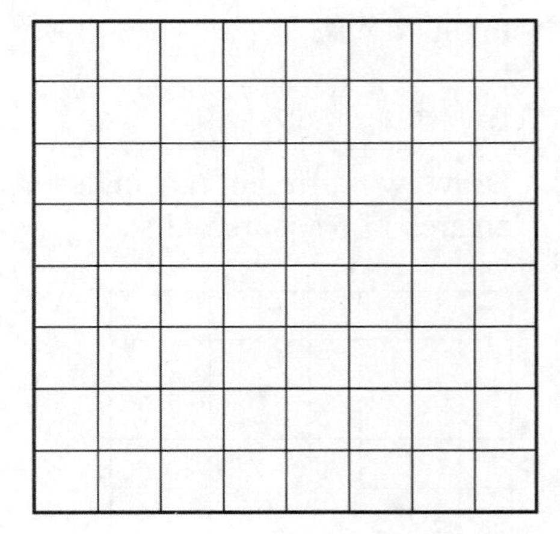

Name: ______________________________ Date: _______________
Instructor: ___________________________ Section: _______________

Introduction to Geometry
8.6 Finding Area

Vocabulary

square unit • area • perimeter • square inches

1. The ____________ of a figure is the measure of the surface inside the figure, which is measured in square units.

Step-by-Step Video Notes

Watch the Step-by-Step Video lesson and complete the examples below.

Example	Notes
1. Use the formula for area of a rectangle to find the area of the rectangular photo shown below. 9 inches 6 inches $A = lw$, where l is length and w is width $A = 9 \text{ in.}(\square \text{ in.}) = \square$ square inches Answer:	
2. Use the formula for area of a square to find the area of a square deck with sides 3 meters. 3 m 3 m $A = s^2$, where s is the side $A = (\square \text{ meters})^2 = \square$ square meters Answer:	

Example	Notes
3. Use the formula for area of a triangle to find the area of a right triangle with a base of 4 cm and a height of 3 cm. 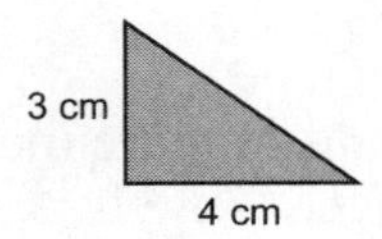 Answer:	
4. Find the area of the infield of a major league baseball diamond which is a square whose sides are 90 feet long. 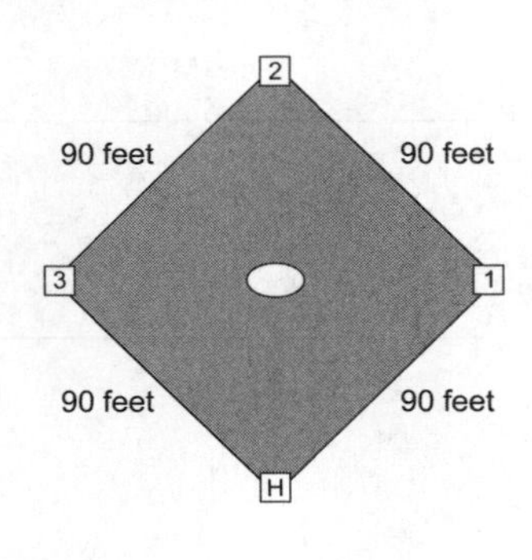 Answer:	

Helpful Hints

The basic shapes have formulas which can be used to find their area.

Remember to include the square units of length when finding area of a shape.

Concept Check

1. What are the formulas for finding the area of a rectangle, a square, and a right triangle?

Practice

Find the area of the following.

2. A rectangle whose length is 12 inches and width is 3 inches
3. A square whose sides are 7 meters long
4. A right triangle with a base of 12 cm and a height of 5 cm
5. A diamond logo on a shirt which is a square whose sides are 8 cm long.

Name: ______________________________ Date: ________________
Instructor: ____________________________ Section: ______________

Introduction to Geometry
8.7 Understanding Circles

Vocabulary
circle • diameter • perimeter • radius • circumference • pi

1. A ____________ is a figure in which all points on the circle are the same distance from a fixed point called the center.

2. The ____________ of a circle is the distance from the center to a point on the circle.

3. The ____________ of a circle is the distance around the circle.

Step-by-Step Video Notes
Watch the Step-by-Step Video lesson and complete the examples below.

Example	**Notes**
1. Use the appropriate formula to find the radius of a pizza with a diameter of 10 inches. Use the formula $r = \frac{1}{2}d$, where d is the diameter. $r = \frac{1}{2}(\square \text{ inches}) = \square \text{ inches}$ Answer:	
2. A neighbor purchased a 14-foot circular trampoline for his children. Does this situation involve the diameter or the radius? 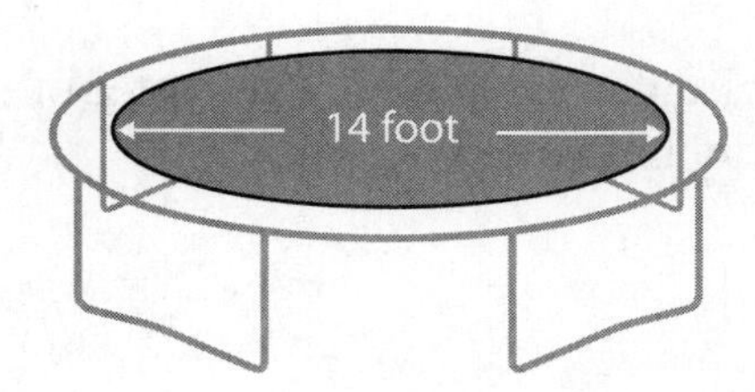 Review the definitions for diameter and radius. Answer:	

Example	**Notes**
3. In softball, the circle around the pitcher's mound is drawn 6 feet from where the pitcher stands. Does this situation involve the diameter or the radius? 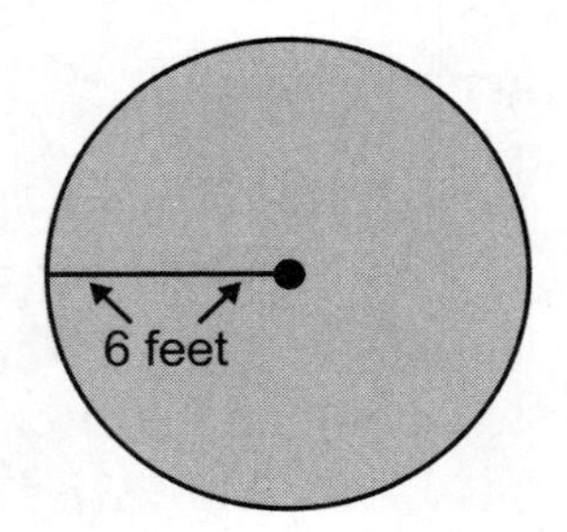 Answer:	

Helpful Hints
The radius is always smaller than the diameter; the radius is half of the diameter.

The radius extends from the center of a circle, while the diameter passes through the center.

Concept Check
1. Define these parts of a circle: radius, diameter and circumference.

Practice
Does the situation described involve the diameter of the radius?

2. A toy helicopter has moveable blades that are 5 inches long.

3. A child's circular board game folds along the center with a measure of 15 inches.

Find the radius in the following situations.

4. A circular garden with a diameter of 4 feet.

5. A circular drum with a diameter of 12 inches.

Name: ______________________________ Date: ________________
Instructor: ____________________________ Section: ______________

Introduction to Geometry
8.8 Finding Circumference

Vocabulary
diameter • perimeter • radius • circumference

1. $\pi = \frac{C}{d}$, where C is the circumference and d is the ____________.

2. The ____________ of a circle is the distance around the circle.

Step-by-Step Video Notes
Watch the Step-by-Step Video lesson and complete the examples below.

Example	Notes
1. Find the circumference of a circle with a radius of 8 inches. Find the exact answer in terms of π. Then find the approximation using 3.14 for π. Use the formula $C = 2\pi r$ $C = 2\pi(\square \text{ in.}) = \square\ \pi \text{ in.}$ Use 3.14 for π. $C = \square(3.14) \text{ in.} = \square \text{ in.}$ Answer:	
2. Find the circumference of a circle with a diameter of 5 cm. Find the exact answer in terms of π. Then find the approximation using 3.14 for π. $C = \pi(\square \text{ cm}) = \square\ \pi \text{ cm}$ Use 3.14 for π. Answer:	

Example	Notes
3. The earth's equator forms a circle. Estimate the number of miles a ship would have to travel if it went around the earth at the equator. Use 7900 miles as an approximation of the earth's diameter. Use 3.14 for π. Answer:	

Helpful Hints
Exact answers for circumference are left in terms of π.

Approximations for circumference use $\frac{22}{7}$ or 3.14 for the value of π.

Concept Check

1. What are the two formulas for finding the circumference of a circle?

Practice
Find the circumference of the following circles. Find the exact answer in terms of π. Then find the approximation using 3.14 for π.

2. A circle with a radius of 6 meters

3. A circle with a diameter of 7 inches

4. The bottom of an empty farm silo forms a circle. The diameter of the silo is 18 feet.

5. The lid to a trash barrel is a circle of radius 1.5 feet.

Name: ______________________________ Date: ______________

Instructor: ____________________________ Section: ______________

Introduction to Geometry
8.9 Finding Area – Circles

Vocabulary

diameter • area • radius • circumference

1. The ____________ of a figure is the measure of the surface inside the figure.

Step-by-Step Video Notes

Watch the Step-by-Step Video lesson and complete the examples below.

Example	**Notes**
1. Find the area of a circle with a radius of 6 feet. Use the approximate value of 3.14 for π. 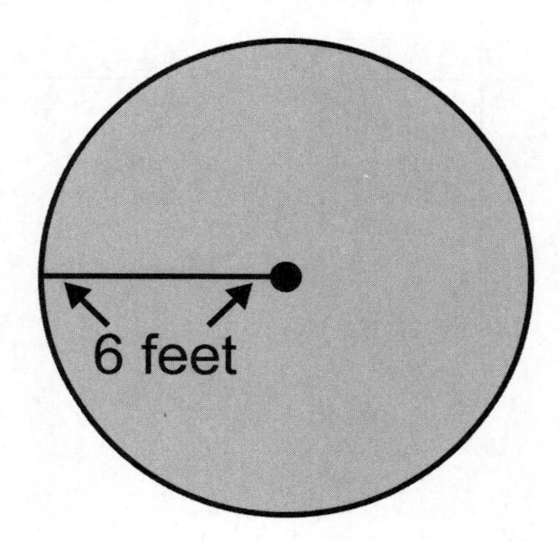 Use the formula $A = \pi r^2$, where A is the area and r is the radius. Enter the radius. $A = (3.14)(\square \text{ feet})^2$ Square the radius. $A = (3.14)(\square \text{ square feet})$ Answer:	

Example	Notes
2. Find the area of a circle with a diameter of 6 meters. Use the approximate value of 3.14 for π. 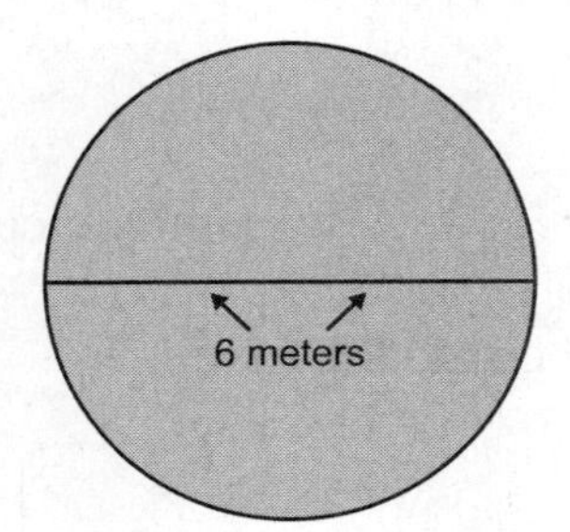 Find the radius using $r = \frac{1}{2}d$ Next use the formula $A = \pi r^2$. Answer:	
3. Find the area of a circle with a diameter of 7 feet. Use the approximate value of $\frac{22}{7}$ for π. Answer:	

Helpful Hints
Remember that area is measured in square units.

If the diameter is given when finding the area of a circle, it must be divided by 2 to get the radius.

Concept Check
1. What is the formula for finding the area of a circle?

Practice
Find the area of the following circles. Use the approximate value of 3.14 for π.
2. A circle with a radius of 10 feet
3. A circle with a radius of 2.5 meters
4. A circle with a diameter of 3.5 cm
5. A circle with a diameter of 10 mi

Name: ______________________________ Date: ________________
Instructor: ____________________________ Section: ______________

More Geometry
9.1 Volume – Definitions and Units

Vocabulary

Volume • cube • box • area

1. A ____________ is a rectangular box in which every side is a square.

2. The ____________ of a three-dimensional figure is the amount of space inside the figure.

Step-by-Step Video Notes

Watch the Step-by-Step Video lesson and complete the examples below.

Example	Notes
1. Find the volume of a box with a length of 4 inches, a width of 3 inches, and a height of 4 inches.  How many cubes will it take to fill the bottom layer? $4 \times 3 = \square$ How many layers will fill the box? $\square$ Multiply the number of layers by the number of cubes in each layer. Answer in cubic units. Answer:	

Example	Notes
2. Find the volume of a cube that measures 4 cm on each side. 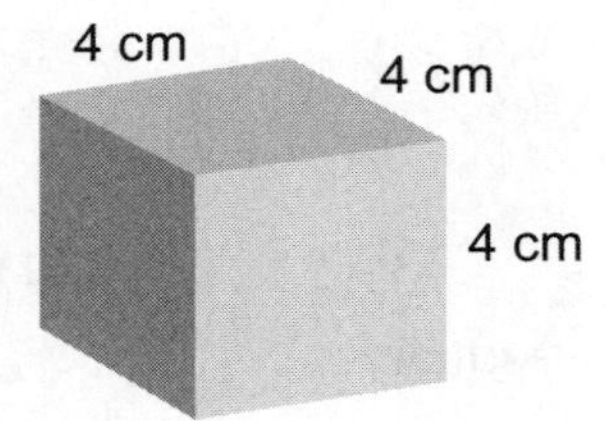 How many 1 cm cubes will it take to fill the bottom layer? ☐ Answer:	
3. Find the volume of a box with length 5 m, width 1 m, and height 3 m. 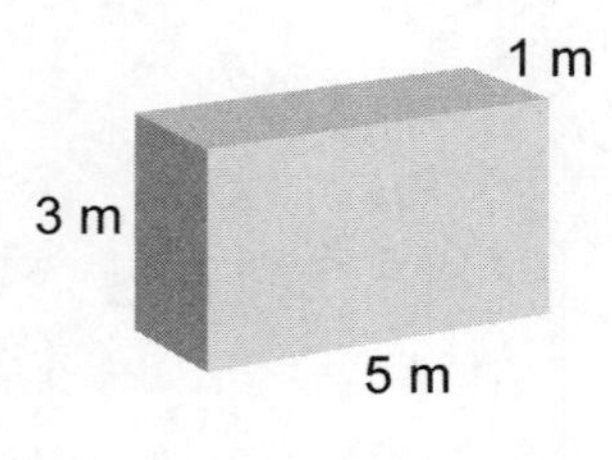 Answer:	

Helpful Hints
Volume involves working with three dimensions; length, width, and height.

Volume is measured in cubic units. You can write out the cubic units for the label (cubic meters), or use the exponent and the abbreviation (m^3).

Concept Check

1. How does measuring area help to find volume for a cube or a box?

Practice

Find the volume of a cube with the given side measure.

2. 6 in.

3. 9 feet

Find the volume of a rectangular box with the given dimensions.

4. length 4 ft, width 2 ft, height 7 ft

5. length of 3 m, width of 6 m, height of 5 m

Name: ______________________________ Date: ________________
Instructor: ___________________________ Section: ______________

More Geometry
9.2 Finding Volume

Vocabulary

cylinder • sphere • pi • formula • radius

1. Using a _____________ is a faster and more practical way to find the volume of a three-dimensional object than by counting unit cubes.

Step-by-Step Video Notes

Watch the Step-by-Step Video lesson and complete the examples below.

Example	Notes
1. Use the formula to find the volume of a rectangular box with a length of 12 inches, a width of 10 inches, and a height of 5 inches. 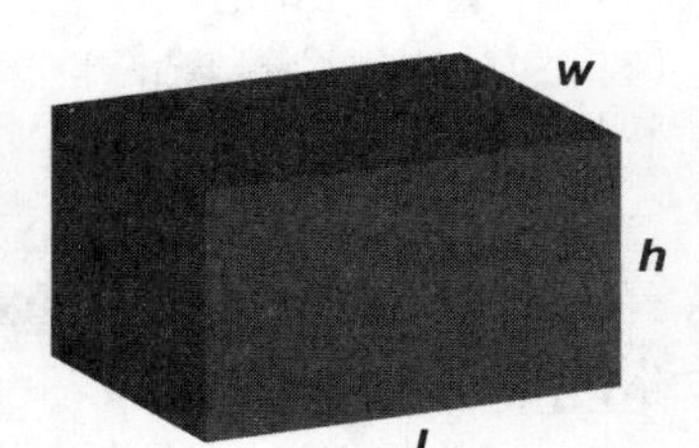 $V = lwh = 12 \times 10 \times 5 = \square$ Answer:	
2. Use the formula to find the volume of a cube that measures 2.5 km on each side. Round to the nearest hundredth. 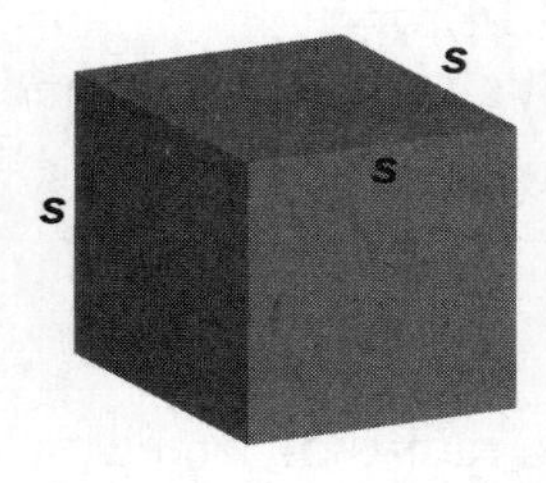 $V = s^3 = (\square)^3$ Answer:	

Example	Notes
3. Use $V = \pi r^2 h$ and the approximate value of 3.14 for π to find the volume of a cylinder with radius of 3 inches and a height of 5 inches. 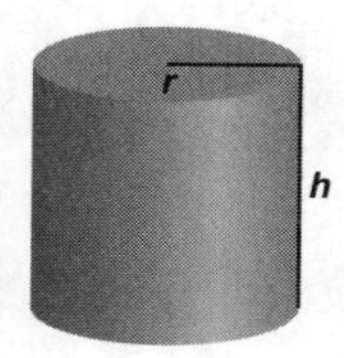 Answer:	
4. Use the formula and the approximate value of 3.14 for π to find the volume of a sphere with a radius of 6 mm. 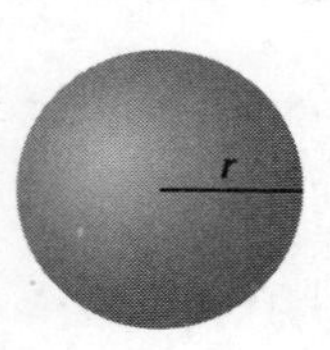 Answer:	

Helpful Hints

Use the formulas $V = lwh$, $V = s^3$, $V = \pi r^2 h$, and $V = \frac{4\pi r^3}{3}$ to find the volumes of rectangular boxes, cubes, cylinders, and spheres, respectively.

Concept Check

1. How does using a formula make it simpler to find the volume of an object?

Practice

Find the volume of the figure with the given dimensions.

2. a cube with a side length 8 ft

3. a rectangular box with length 8 cm, width 12 cm, height 11 cm

Find the volume of the figure with the given dimensions. Use 3.14 for π.

4. a cylinder with a radius of 8 in. and a height of 9 in.

5. a sphere with a radius of 1 meter

Name: ______________________________ Date: ______________
Instructor: ____________________________ Section: ______________

U.S. and Metric Measurement
9.3 Square Roots

Vocabulary
perfect square • square root • area of a square • radical

1. If $a^2 = b$, then a is the ____________ of b.

2. The symbol which denotes square root, $\sqrt{}$, is called the ____________ sign.

Step-by-Step Video Notes
Watch the Step-by-Step Video lesson and complete the examples below.

Example	Notes
1. Find the square root of 36. When you multiply two identical factors to result in another number, the square root is one of those identical factors. The square root of 36 is $\square$, because $\square^2 = 36$. Answer:	
2. Simplify $\sqrt{64}$. Find two identical factors whose product is 64. The square root of 64 is $\square$, because $\square^2 = 64$. Simplify $\sqrt{400}$. Find two identical factors whose product is 400. The square root of 400 is $\square$, because $\square^2 = 400$.	

Example	Notes
3. The area of a square is 49 cm^2. What is the length of each side? (square with sides labeled s and s) The area of a square is found by multiplying the side length by itself. Since $\square \times \square = 49$, $\sqrt{49} = \square$. Answer:	
4. Approximate $\sqrt{75}$ by finding the two consecutive whole numbers the square root lies between. The perfect square just less than 75 is 64, and $\sqrt{64} = \square$. The perfect square just greater than 75 is $\square$, and $\sqrt{\square} = \square$. Therefore, $\sqrt{75}$ is between $\square$ and $\square$. Answer:	

Helpful Hints

A perfect square is a number that has a whole number square root. If a whole number is not a perfect square, use a calculator, or approximate the square root by finding two perfect squares that whole number lies between.

Concept Check

1. Can you find the perimeter of a square if you know its area?

Practice

Find the square root of the given number.

2. 36

3. 144

Use a calculator to approximate each square root to the nearest hundredth.

4. $\sqrt{154}$

5. $\sqrt{55}$

Name: ______________________________ Date: ________________
Instructor: ____________________________ Section: ______________

U.S. and Metric Measurement
9.4 The Pythagorean Theorem

Vocabulary
Pythagorean Theorem • hypotenuse • leg • diagonal

1. The longest side of a right triangle, opposite the right angle, is the ____________.

2. The ____________ states that in a right triangle, the sum of the squares of the legs is equal to the square of the hypotenuse, or $\text{leg}^2 + \text{leg}^2 = \text{hypotenuse}^2$.

Step-by-Step Video Notes
Watch the Step-by-Step Video lesson and complete the examples below.

Example	Notes
1. Find the length of the hypotenuse of a right triangle with legs that measure 3 cm and 4 cm. To find the hypotenuse, use the formula $\text{hypotenuse} = \sqrt{\text{leg}^2 + \text{leg}^2}$. $\text{hypotenuse} = \sqrt{3^2 + 4^2} = \sqrt{9 + \square} = \sqrt{\square} = \square$ Answer:	
2. One leg of a right triangle measures 9 inches, and the hypotenuse measures 15 inches. Find the length of the other leg. To find the length of the missing leg, use the formula $\text{leg} = \sqrt{\text{hypotenuse}^2 - \text{leg}^2}$. Substitute for the hypotenuse and leg. $\text{leg} = \sqrt{15^2 - \square^2} = \sqrt{\square}$ Answer:	

Example	Notes
3. Find the missing side of the right triangle. Round to the nearest tenth. 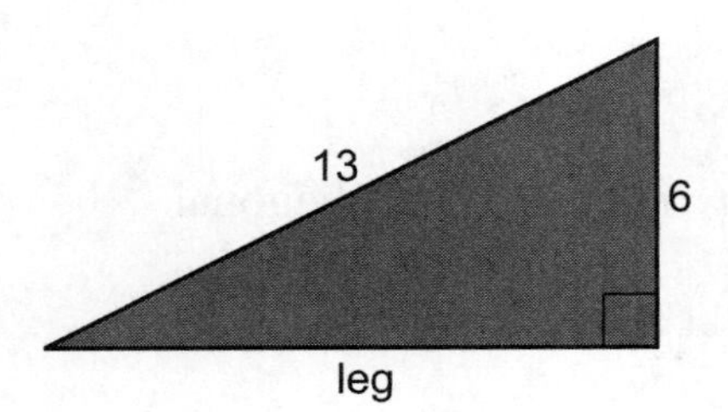 Identify the formula to use. Substitute the given values into the formula. Find the missing side. Answer:	
4. A standard computer monitor has a length of 20 inches and a width of 15 inches. What is the measure of the diagonal of the monitor? The monitor is a rectangle. The diagonal divides the rectangle into 2 identical right triangles. The length and width are the legs, and the diagonal is the hypotenuse. Identify the formula to use. Answer:	

Helpful Hints

The Pythagorean Theorem works with right triangles or the many real-life situations where a right triangle can be applied to help find a distance. Be sure to identify whether the side of the triangle you are trying to find is a leg or the hypotenuse, and use the appropriate formula. Quite often, the Pythagorean Theorem is stated as $a^2 + b^2 = c^2$.

Concept Check

1. How do you find the length of the diagonal of a rectangle?

Practice

Find the hypotenuse of a right triangle with the given length of the legs.

2. Leg = 5 ft Leg = 12 ft

3. Leg = 16 m Leg = 12 m

Find the length of the missing leg of a right triangle given the hypotenuse and a leg.

4. Leg = 24 in. Hypotenuse = 25 in.

5. Leg = 24 cm Hypotenuse = 26 cm

Name: ____________________________ Date: ______________
Instructor: __________________________ Section: ______________

U.S. and Metric Measurement
9.5 Similar Figures

Vocabulary

similar figures • proportion • corresponding sides • ratio

1. ____________ have the same shape and corresponding angles with the same measure.

Step-by-Step Video Notes

Watch the Step-by-Step Video lesson and complete the examples below.

Example	Notes
1. Identify the corresponding sides and set up proportions. Determine if the figures are similar. 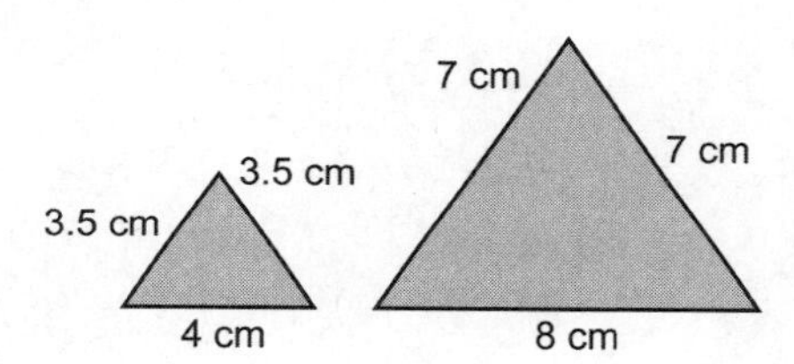 The left and right sides of the smaller triangle measure 3.5 cm. The corresponding sides of the larger triangle measure 7 cm. The ratio is $\frac{3.5}{7}$. Compare the bottom side of the smaller triangle to the bottom side of the larger triangle. Is $\frac{3.5}{7} = \frac{\square}{\square}$?	
2. Identify the corresponding sides and set up proportions. Determine if the figures are similar. 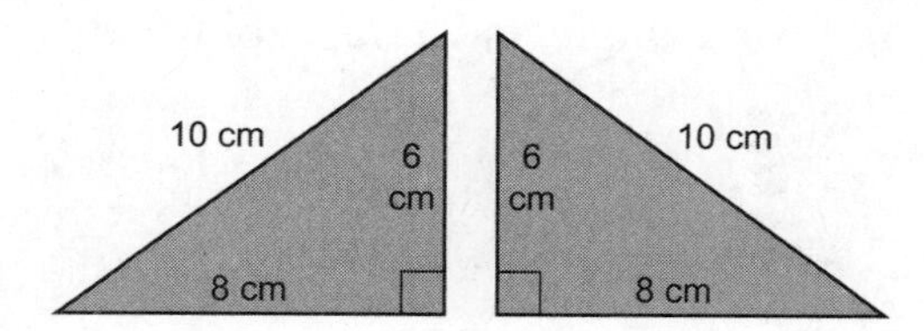 Both figures are right triangles. $\frac{6}{6} = \frac{\square}{\square} = \frac{\square}{\square}$, so the figures ____________ similar.	

Example	Notes
3. Determine if the figures in each pair are similar.	

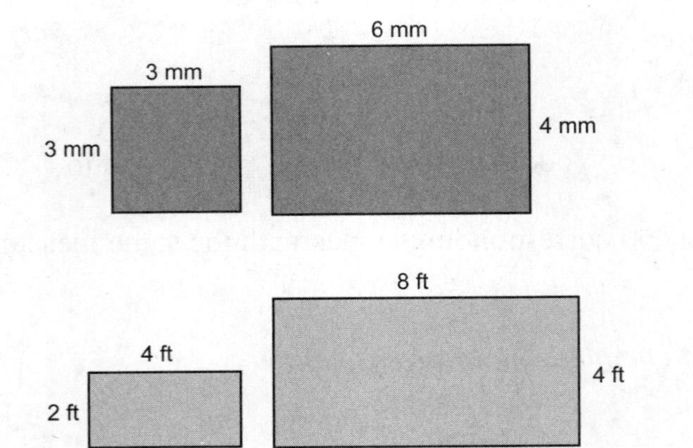

In the first set of rectangles, is $\frac{3}{4} = \frac{3}{6}$?

In the second set of rectangles, is $\frac{2}{4} = \frac{\square}{8}$?

Answer:

Helpful Hints

The concept of similar figures applies to many everyday applications such as photography, art, scale models, and maps. If figures are similar, they are proportional, or "to scale."

If the corresponding sides of two triangles are proportional, the triangles are similar. Likewise, two rectangles are similar if their corresponding sides are proportional.

Concept Check

1. Can you think of geometric shapes other than squares that are always similar?

Practice

Determine if the figures are similar.

2. Triangle A with sides 4 m, 5 m, 6m
   Triangle B with sides 8 m, 12 m, 10 m

3. Triangle C with sides 9 ft, 9 ft, 7 ft
   Equilateral Triangle D with sides 9 ft

Can the following pairs of figures be similar?

4. A rectangle and a trapezoid

5. A very small square measured in mm and a very large square measured in km

Name: ______________________________ Date: ________________
Instructor: ____________________________ Section: ______________

U.S. and Metric Measurement
9.6 Finding Unknown Lengths

Vocabulary
corresponding sides • proportion • proportional • unknown

1. Corresponding sides of similar figures are ____________.

Step-by-Step Video Notes
Watch the Step-by-Step Video lesson and complete the examples below.

Example	Notes
1. Write a proportion for the corresponding sides of the similar figures.	

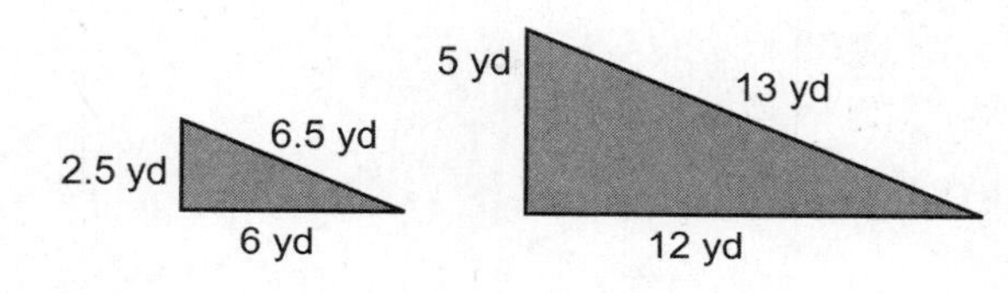

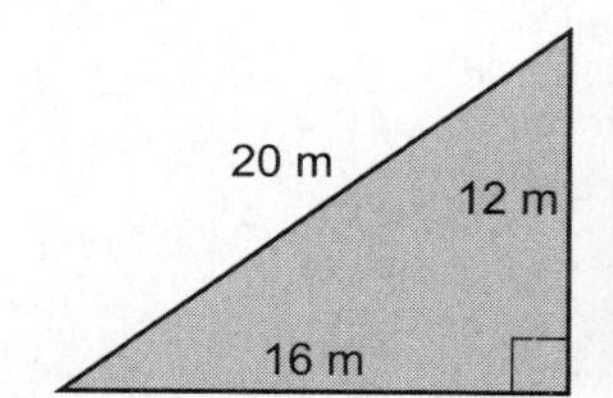

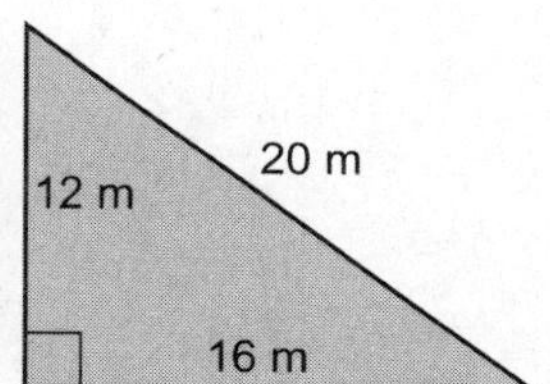

The proportion for the corresponding sides of the first pair of triangles is $\frac{2.5}{5} = \frac{\square}{12} = \frac{6.5}{\square}$.

Each ratio in the proportion is equal to $\frac{1}{\square}$.

In the second pair, the figures have exactly the same shape, therefore they are ____________.

Answer:

Example	Notes
2. Find the unknown side length in the similar figures. 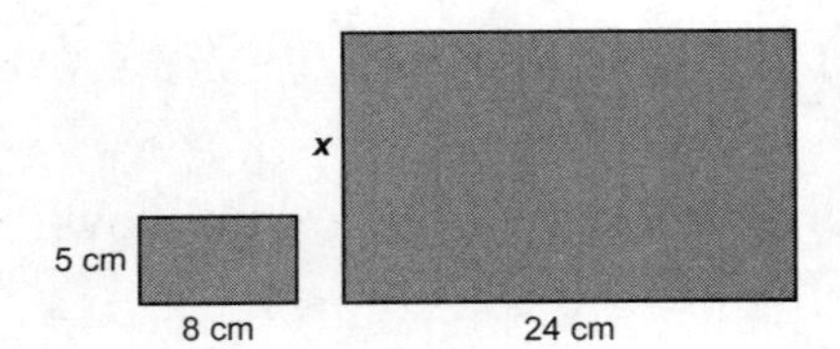 The proportion is $\frac{\square}{x} = \frac{8}{\square}$. Answer:	
3. Find the unknown side length in the similar figures. 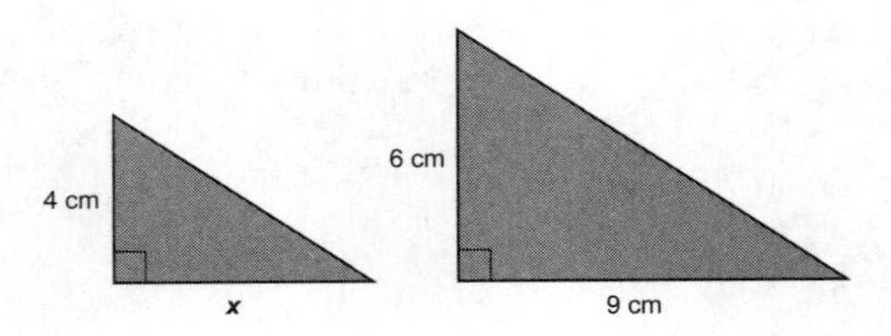 The proportion is $\frac{4}{\square} = \frac{x}{\square}$. Answer:	

Helpful Hints

If figures are similar, the ratios of the corresponding sides will be equal. Line up the corresponding sides of similar figures in a proportion to find unknown lengths.

Concept Check

1. Is a 30 in. wide by 36 in. tall poster similar to a 4 in. by 6 in. photo?

Practice

Solve for x.

2. A 6 in. by 9 in. rectangle is similar to a 24 in. by x in. rectangle

3. A triangle with sides 6 m, 10 m, 8 m is similar to a triangle with sides 9 m, 15 m, and x m.

Find the unknown length.

4. A 5 cm by 12 cm rectangle is similar to a 15 mm by $\square$ mm rectangle.

5. $\frac{8 \text{ ft}}{12 \text{ ft}} = \frac{10 \text{ ft}}{\square}$

Name: ______________________________ Date: ______________
Instructor: ___________________________ Section: ______________

Statistics
10.1 Bar Graphs and Histograms

Vocabulary
bar graph • double bar graph • measurement • frequency distribution table • histogram

1. A ____________ is a special type of bar graph where the width of the bar represents an interval, or range of numbers.

Step-by-Step Video Notes
Watch the Step-by-Step Video lesson and complete the examples below.

Example	Notes
1. Use the bar graph below to answer the question. How many Country Music Association Awards did Loretta Lynn win? 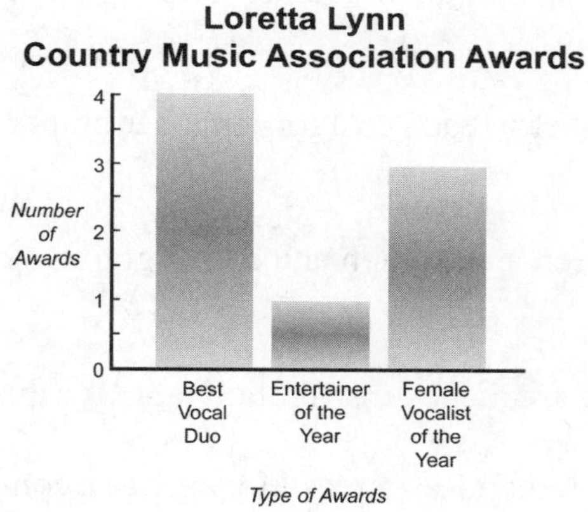 **Loretta Lynn Country Music Association Awards** Number of Awards: 0, 1, 2, 3, 4 Best Vocal Duo; Entertainer of the Year; Female Vocalist of the Year *Type of Awards* The height of each bar represents the number in each category. Enter these numbers from the graph. Number of awards = 4 + ☐ + ☐ Answer:	
2. Use the bar graph in example #1 to answer the question. In which category did she win the fewest awards? Which bar is the lowest? ____________	

Example	Notes
3. In an Intro to Music class, letter grades are distributed based on the following scale: score of 90-100 an A, 80-89 a B, 70-79 a C, 60-69 a D, and below 60 is an F. Enter the tally and frequency in the table below for a class with scores of 78, 69, 82, 95, 92, 80, 47, 89, 81, and 99.	

Intro To Music Grade Distribution

Grade Intervals	Tally	Frequency
90 – 100		
80 – 89		
70 – 79		
60 – 69		
Below 60		

Helpful Hints

A bar graph can be used to display data over time, to compare amounts, or show how often a particular amount will occur.

The height of the rectangular bar indicates the number in each category in a bar graph.

Concept Check

1. Name two similarities and one difference between a bar graph and a histogram.

Practice

Use the bar graph below to answer the following.

2. How many softball team members voted?

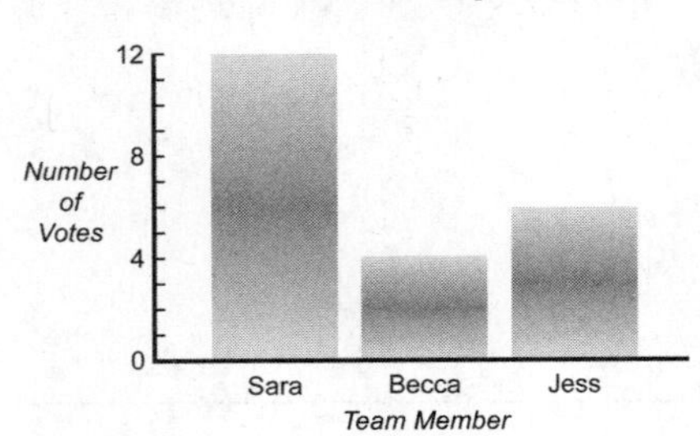

3. Which player received the fewest votes?

Fill in the frequency distribution table for the given data below.

4. The Human Resources Manager at a company tracks employee sick days in the categories of 0-2, 3-5, 6-8, 9-11, and 12 or more. The art department employees used sick days of 4, 6, 2, 3, 9, 0, 1, 0, 3, 2, 4, 7, 0, 1, and 15.

Employee Sick Days Distribution

# of days used	Tally	Frequency
0 – 2		
3 – 5		
6 – 8		
9 – 11		
12 or more		

Name: ______________________________ Date: ______________
Instructor: ____________________________ Section: ____________

Statistics
10.2 Line Graphs

Vocabulary
double line graphs • line graphs • data points • histograms

1. ____________ can display data, show trends or patterns in data over time, show how often a particular data value occurs, compare two or more types of data, and use data points that are connected with straight line segments.

Step-by-Step Video Notes
Watch the Step-by-Step Video lesson and complete the examples below.

Example	Notes
1. Use the line graph below to answer the question. Has the average life expectancy of humans increased or decreased since 1920? 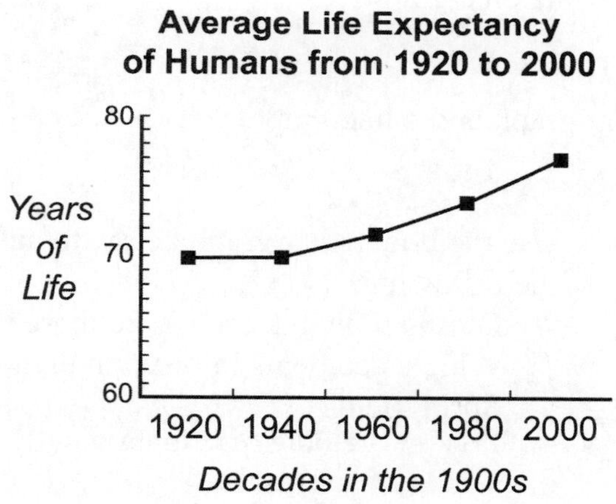 **Average Life Expectancy of Humans from 1920 to 2000** Years of Life: 60, 70, 80 1920 1940 1960 1980 2000 *Decades in the 1900s* Answer:	
2. Use the line graph in example #1 to answer the question. During which 20-year interval did the life expectancy increase the most? Answer:	

Example

Notes

3. Use the double line graph below to answer the question. Which countries had the same life expectancy in for 2000 and 1998?

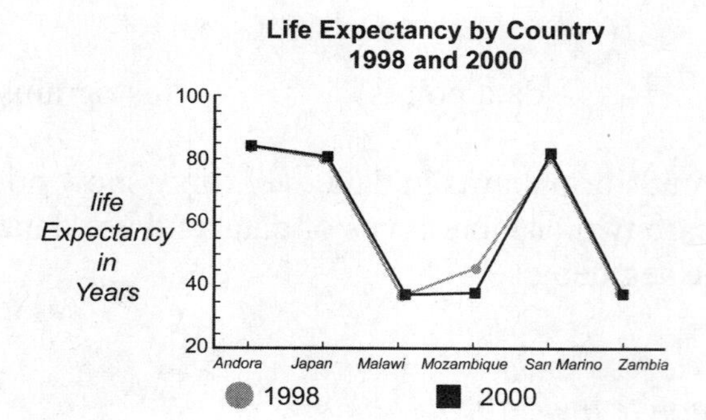

Helpful Hints

A line graph is usually used to show trends or patterns of data over time, while a double line graph is used to compare more than one set of data on a graph.

Concept Check

1. State at least two differences between a bar graph and a line graph.

Practice

Use the bar graph below to answer the following.

2. Have the vehicle accidents in Springfield increased or decreased since 1960?

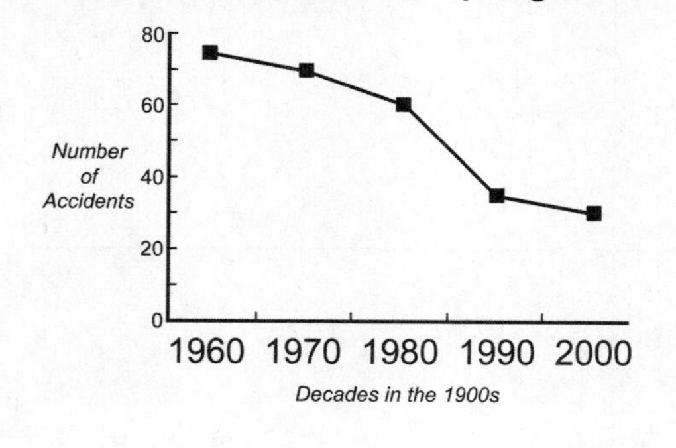

3. During which decade was the decrease in accidents the most?

Use the double bar graph below to answer the following.

4. During which decade were there more vehicle accidents in Newton than in Springfield?

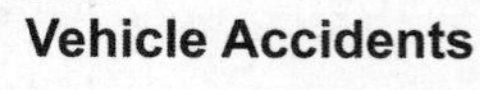

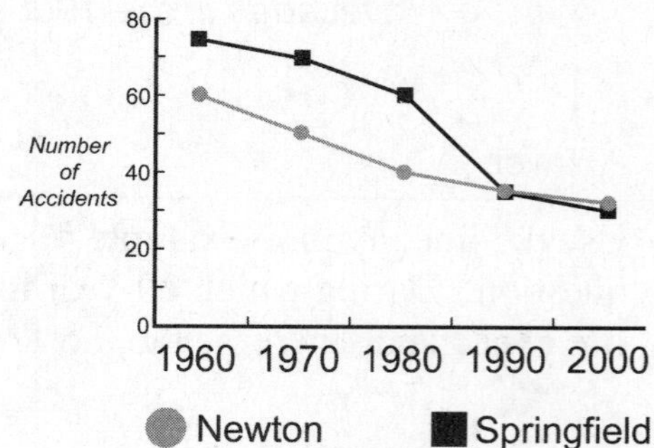

Name: ______________________________ Date: ________________
Instructor: ___________________________ Section: ______________

Statistics
10.3 Circle Graphs

Vocabulary

line graph • circle graph • amounts • percents

1. A ______________, also called a pie chart, is drawn to show actual amounts or percents and is used to show how one set of data is divided.

Step-by-Step Video Notes

Watch the Step-by-Step Video lesson and complete the examples below.

Example	Notes
1. Use the circle graph below to answer the question. What does Nikeshia plan to spend the most money on? **Average Costs of School** Books $400; Tuition $1200; Housing $800; Food $800; Other $400 Answer:	
2. Use the circle graph in example #1 to answer the question. What total amount does Nikeshia plan to spend on food and housing? Find the sum of the amount from the food category and the housing category. Answer:	

Example	Notes
3. Use the circle graph below to answer the question. Which assignment type counts the most toward the student's final grade?	

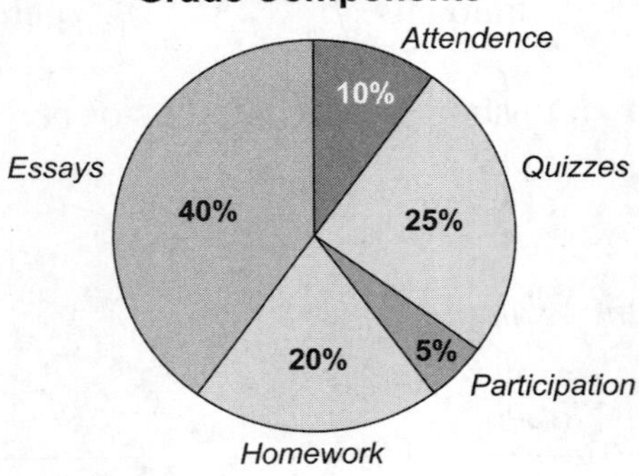

Answer:

Helpful Hints
Circle graphs are most often used to display relationships that compare parts to a whole.

In circle graphs the sum of all of the parts must equal the total amount or 100%.

Concept Check
1. What is different about a circle graph when compared to a bar graph or line graph?

Practice
Use the circle graphs to answer the following questions.

2. What eye color is the least frequent in this class of students?

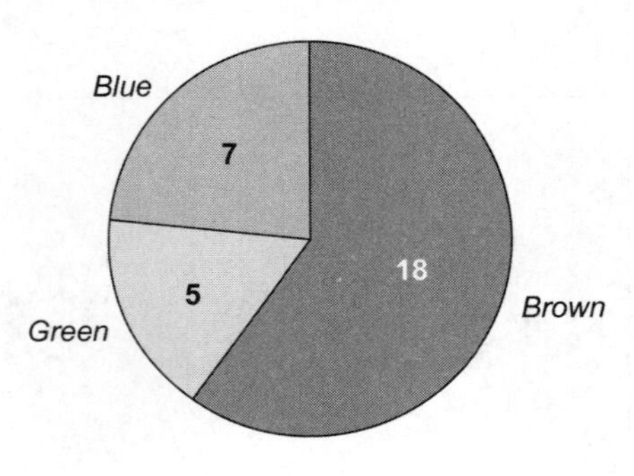

3. How many students are in this class?

4. What percentage of students in this class has blue eyes?

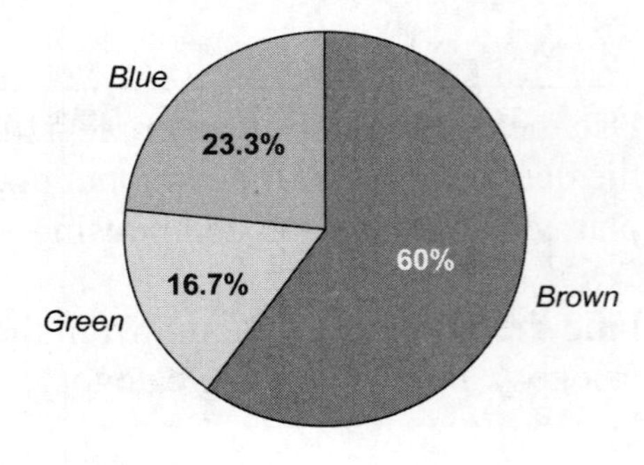

Name: ______________________________ Date: ________________
Instructor: ___________________________ Section: ______________

Statistics
10.4 Mean

Vocabulary
measures of central tendency • statistics • mean • data

1. ____________ is information about a group or topic, often consisting of a set of numbers.

2. The ____________ is the sum of the values in the list of numbers divided by the number of values.

Step-by-Step Video Notes
Watch the Step-by-Step Video lesson and complete the examples below.

Example	Notes
1. Calculate the mean for the following list of numbers. 3, 5, 7, 2, 11, 6, 8, 2, 10 Find the sum of all of the values in the list. $3+5+\square+2+11+6+8+\square+10=\square$ There are 9 values in the list. Divide the sum by the number of values in the list. $\square \div 9=\square$ Answer:	
2. Calculate the mean of the following list of test scores. 93, 86, 95, 98, 82 Find the sum. $\square$ There are $\square$ values in the list. Divide the sum by the number of values in the list. Answer:	

Example	Notes
3. Eli saved a portion of his allowance for 6 weeks. He saves $14 the first week, $12 the second week, $6 the third week, and $4, $10, and $2 for the remaining weeks. On average, how much did he save each week? Find the average, which is the same as the mean, saved over the six weeks.	

Helpful Hints

Mean is a measure of central tendency because it represents an entire list of numbers as well as the "center" of the data.

The mean is often referred to as the average.

Concept Check

1. Describe how to find the mean of a list of numbers.

Practice

Find the mean of each list of numbers.

2. 10, 8, 4, 7, 5, 8, 10, 4

3. 55, 42, 33, 61, 30

4. The five players on Jamison's basketball team collected donations for new team uniforms. Jamison collected $18. His teammates collected $12, $34, $26 and $10. On average, how much did each player collect?

Name: ______________________________ Date: ________________
Instructor: ___________________________ Section: _____________

Statistics
10.5 Median

Vocabulary
median • statistics • mean • data

1. ____________ is the study and use of data.

2. The ____________ of a set of ordered numbers is the middle number.

Step-by-Step Video Notes
Watch the Step-by-Step Video lesson and complete the examples below.

Example	Notes
1. Find the median of the following list of numbers. 3, 5, 7, 2, 11, 6, 8, 2, 10 Write the numbers in order from smallest to largest. 2, □, 3, 5, □, □, 8, 10, □ Does the list contain an odd number or even number of values? Odd; there are 9 values. Mark off one number from the beginning of the list and from the end of the list. ~~2~~, 2, 3, 5, □, □, 8, 10, ~~11~~ Repeat this step. ~~2~~, ~~2~~, 3, 5, □, □, 8, ~~10~~, ~~11~~ Repeat this step until middle number is reached. Answer:	

Example	Notes
2. Find the median of the following list of numbers. 44, 32, 31, 56, 77, 65 Order the numbers. Since there are an even number of values, the median is the average of the two middle numbers. Find this average. $44 + \square = \square \qquad \square \div 2 = \square$ Answer:	
3. A poll of nine students shows the numbers of Facebook friends they have are 395, 486, 166, 430, 172, 159, 723, 582, and 319. Find the median number of Facebook friends for these nine students. Answer:	

Helpful Hints
Remember that the values must be ordered before finding the median of a list of numbers.

Concept Check

1. How do you find the median of a list that has an even number of values? How does this differ from the process of finding the median of list that has an odd number of values?

Practice
Find the median of each list of numbers.

2. 10, 8, 4, 7, 5, 8, 10

3. 55, 43, 33, 61, 30, 41

4. Sarah's math test scores for the first term were 95, 92, 81, 98, 84, 86, and 94. Find the median of Sarah's math test scores.

Name: ______________________________ Date: ________________
Instructor: ___________________________ Section: ______________

Statistics
10.6 Mode

Vocabulary

median • statistics • data • mode

1. The ____________ of a list of numbers is the number that occurs most often in the list.

Step-by-Step Video Notes

Watch the Step-by-Step Video lesson and complete the examples below.

Example	Notes
1. Find the mode(s) of the following list of numbers. 3, 5, 7, 2, 11, 6, 8, 2, 10 Write the numbers in order from smallest to largest. 2, 2, 3, □, 6, □, □, 10, 11 Choose the number(s) that occur the most often. Answer:	
2. Find the mode(s) of the following list of numbers. 44, 32, 31, 56, 77, 65, 65, 44, 65 Order the numbers. 31, □, 44, □, □, □, 65, 77 Answer:	
3. Find the mode(s) of the following list of numbers. 325, 612, 367, 692, 451 Note that a list of numbers can have more than one mode or no mode. Answer:	

Example	Notes
4. According to the National Weather Service, the high temperatures (in °F) for the first two weeks of June 2010 were recorded as 105, 106, 109, 112, 116, 119, 119, 115, 111, 107, 106, 103, 104, and 110. Find the mode(s) of the high temperatures for this period. Answer:	

Helpful Hints

It is not a required step to order the numbers when finding the mode, but it can be helpful.

The mode is not just any number that occurs more than once, it is the number that occurs more often than the rest.

Remember that there can be one mode, more than one mode, or no mode.

Concept Check

1. How is the mode different than the other two measures of central tendency, the mean and the median?

Practice

Find the mode(s) of each list of numbers.

2. 65, 35, 40, 55, 40, 30, 70

3. 10, 8, 4, 7, 5, 8, 10

4. The students in a kindergarten class each have a box for storing crayons and pencils. The number of crayons in the boxes is 8, 6, 7, 6, 2, 5, 7, 4, 3, and 8. Find the mode(s) of the crayons in the boxes.

Name: ______________________________ Date: ______________

Instructor: ___________________________ Section: ______________

Signed Numbers
11.1 Signed Numbers and Absolute Value

Vocabulary

positive • negative • opposites • absolute value

1. The ____________ of a number is its distance from zero on a number line.

2. ____________ are two numbers that are the same distance from zero but appear on different sides of zero on the number line.

Step-by-Step Video Notes

Watch the Step-by-Step Video lesson and complete the examples below.

Example	Notes
1. Insert a $<$ or $>$ in the blank to make each statement true. $4 \square -4 \qquad -8 \square -15$ When graphed on a number line, numbers will appear in order from smallest to largest as the number line goes from left to right. In this case, any positive number is always greater than any negative number, so $4 \square -4$. When placed on a number line, -8 is to the ____________ of -15, so $-8 \square -15$. Answer:	
2. Write "The opposite of -25" in symbols. Find the opposite and simplify. In symbols, "$-$" means opposite, so $-(-25)$ means the opposite of -25. $-(-25) = \square$ Answer:	

Example	Notes
3. Find the opposite of (-38). To find the opposite of a number, ___________ its sign. Answer:	
4. Find the absolute value. $\|-13\|$ Absolute value is the distance from zero on a number line. It can never be negative. $\|-13\| = \square$ Answer:	

Helpful Hints

You must use the "−" sign with a negative number to indicate that the number's value is less than 0. You may use, but do not need to use, the "+" to indicate that a positive number's value is more than 0.

If a number is positive its opposite is negative. If a number is negative its opposite is positive. The opposite of 0 is 0.

Absolute value bars work like parentheses. Do what is inside the bars before evaluating the absolute value of the result.

Concept Check

1. Why is it impossible for an absolute value to be negative?

Practice

Which is larger?

2. -88 or -87

3. $-\frac{1}{3}$ or $-\frac{1}{4}$

Find each absolute value.

4. $|-(-24)|$

5. $-|2-4|$

Name: ______________________________ Date: ________________
Instructor: ____________________________ Section: ______________

Signed Numbers
11.2 Adding Signed Numbers

Vocabulary
opposites • additive inverses • absolute value

1. ____________ are also known as additive inverses.

2. The ____________ of a number is its distance from zero on a number line.

Step-by-Step Video Notes
Watch the Step-by-Step Video lesson and complete the examples below.

Example	Notes
1. Add. $-9+(-11)$ To add numbers with the same sign, add the absolute values. The sign of the sum is the same as the sign of the original numbers. $-9+(-11)=\square$ Answer:	
2. Add. $-18+7$ To add two numbers with different signs, subtract the absolute values. The sign of the answer is the same as the sign of the number with the larger absolute value. $18-7=\square$ -18 has a ____________ absolute value than does 7, so the sign is $\square$. Answer:	

Example	Notes
3. Add. $6+(-2.5)$ $6-2.5=\square$ The sign of the answer is $\square$. Answer:	
4. Add $3+(-5)+(-2)$. Add from left to right. $3+(-5)=\square$ $\square+(-2)=\square$ Answer:	

Helpful Hints

The sum of two positive numbers is always positive, and the sum of two negative numbers is always negative.

When adding numbers with different signs, it can be helpful to determine the sign of the answer before you simplify.

Concept Check

1. If adding three or more numbers with different signs, can you first add all of the positive numbers together, then add all of the negative numbers, then apply the procedure for adding numbers with different signs?

Practice

Add.

2. $-8+(-7)$

3. $-\frac{4}{9}+\left(-\frac{7}{9}\right)$

4. $-21+19$

5. $34.2+(-34.2)$

Name: ______________________________ Date: ________________
Instructor: ___________________________ Section: ______________

Signed Numbers
11.3 Subtracting Signed Numbers

Vocabulary

adding the opposite • additive inverses • absolute value

1. Rewriting $-9-(-11)$ as $-9+(+11)$ is an example of ______________ as a way to subtract signed numbers.

Step-by-Step Video Notes

Watch the Step-by-Step Video lesson and complete the examples below.

Example	Notes
1. Subtract $-7-3$. To subtract, change the minus sign to a plus sign and then change the sign of the number being subtracted to its opposite. $-7+(\square)=\square$ Answer:	
2. Subtract $-\frac{3}{4}-\left(-\frac{5}{4}\right)$. Use the Leave, Change, Change method. Leave the first number alone. Change the minus sign to a plus sign. Change the sign of the number being subtracted. $-\frac{3}{4}\square\frac{\square}{\square}$ Add the fractions. $\frac{\square}{\square}$ Simplify the sum. $\frac{\square}{\square}$ Answer:	

Example	Notes
3. Subtract $9-7$, $9-(-7)$, $-9-7$, and $-9-(-7)$. $9-7=\square$ $9-(-7)=\square$ $-9-7=\square$ $-9-(-7)=\square$	
4. Subtract $6-(-8)-4$. Subtract from left to right 2 numbers at a time. $6-(-8)=6+\square=\square$ $\square-4=\square$ Answer:	

Helpful Hints

When subtracting signed numbers, "Leave, Change, Change" or "Add the Opposite" will change each subtraction problem to an addition problem.

Subtracting a negative number is the same as adding a positive number. Subtracting a positive number is the same as adding a negative number.

Concept Check

1. How does the concept of absolute value help to explain the statement "Subtracting a negative number is the same as adding a positive number?" Use $2-(-6)$ as an example.

Practice

Subtract. Use "Add the Opposite."

2. $-8-(-9)$

3. $-\frac{2}{9}-\frac{4}{9}$

Subtract. Use "Leave, Change, Change."

4. $3.1-(-1.9)$

5. $-27.6-(-27.6)$

Name: ______________________________ Date: ______________
Instructor: ___________________________ Section: ______________

Signed Numbers
11.4 Multiplying and Dividing Signed Numbers

Vocabulary
even power • odd power • negative factors • exponent

1. When a negative base is raised to an ______________, the result is always negative.

2. When a negative base is raised to an ______________, the result is always positive.

Step-by-Step Video Notes
Watch the Step-by-Step Video lesson and complete the examples below.

Example	Notes
1. Multiply $(-9)(-11)$. Multiply $9 \cdot 11 = \square$. The factors have ______________ signs. $(-9)(-11) = \square$ Answer:	
2. Multiply $(-1)(-2)(-3)(-4)$. When multiplying an even number of negative factors, the product is ______________. When multiplying an odd number of negative factors, the product is ______________. Multiply. $1 \cdot 2 \cdot 3 \cdot 4 = \square$ There are $\square$ negative factors. $(-1)(-2)(-3)(-4) = \square$ Answer:	

Example	Notes
3. Identify the base and exponent, then evaluate. $(-4)^2$ The base is $\square$, and the exponent is $\square$. $(-4)^2 = (\square)\cdot(\square)$ Answer:	
4. Divide $\frac{-45}{-9}$. Remember that the fraction bar is another way to represent division. Divide. If the two numbers have the same sign, the answer is __________; if they have different signs the answer is __________. $\frac{-45}{-9} = \square$ Answer:	

Helpful Hints

It does not matter if a fraction has a negative sign just in the numerator, just in the denominator, or out front. The result is the same; the value of the fraction is negative.

Concept Check

1. Will the result of $(-4)^3 \div (-8)^2$ be positive or negative?

Practice

Multiply.

2. $(-6)(-5)$

3. $\left(\frac{1}{3}\right)\left(-\frac{3}{5}\right)$

Divide.

4. $32 \div (-1.6)$

5. $\frac{4^3}{(-4)^3}$

Name: ______________________________ Date: ________________
Instructor: ___________________________ Section: ______________

Signed Numbers
11.5 Order of Operations and Signed Numbers

Vocabulary
order of operations • fraction bars • simplify

1. When evaluating an expression with more than one operation, use the ____________.

2. ____________ can also act like parentheses.

Step-by-Step Video Notes
Watch the Step-by-Step Video lesson and complete the examples below.

Example	Notes
1. Simplify $(-5)^2 + 3^2$. First evaluate the expressions with exponents. $(-5)^2 = \square$, and $3^2 = \square$ Now add. $\square + \square = \square$ Answer:	
2. Simplify $(-6+4)^3 \div (-2)$. First evaluate what is inside ____________. $-6+4 = \square$ Evaluate the power. $(\square)^2 = \square$ Divide by (-2). Answer:	

Example	Notes
3. Simplify $(3-6)^2-2(-10)$. $(\square)^2-2(-10)=\square-\square$ Add the opposite. $\square+\square=\square$ Answer:	
4. Simplify $\dfrac{(-4)+6-3}{4^2+(2)(-3)}$. The fraction bar is a way to represent division. To simplify, there are understood parentheses around the numerator and the denominator. Simplify the numerator. $\square$ Simplify the denominator. $\square$ Answer:	

Helpful Hints

Remember that fractions, absolute value bars, { }, and [] act like parentheses when simplifying expressions.

Concept Check

1. To simplify $(-9+6)-2(-5)$, which operation will you perform first?

Practice

Simplify. Use the order of operations.

2. $(-8)-3(-5)$

3. $\left(-\frac{1}{3}\right)(-9)^2$

4. $(-4+2)^3\div(-3+5)^2$

5. $\dfrac{7^2-9\cdot 5}{(-6)^2+5(-7)}$

Name: ______________________________ Date: ______________
Instructor: ___________________________ Section: ____________

Introduction to Algebra
12.1 Variables and Algebraic Expressions

Vocabulary
variable • term • expression • algebraic expression • coefficient

1. A(n) ____________ is a number, variable, or product of numbers and/or variables.

2. A(n) ____________ is a symbol (usually a letter) used to represent an unknown number.

Step-by-Step Video Notes
Watch the Step-by-Step Video lesson and complete the examples below.

Example	Notes
1. Identify the coefficient and the variable in $-4x$. The coefficient, or the numerical part, is $\square$. The variable is $\square$. Answer:	
2. Evaluate the algebraic expression $x+y-5$ for $x=2.5$ and $y=9.1$. Replace the variables with the given values, and simplify using order of operations. $\square+\square-5$ Add. $\square-5$ Subtract. $\square-5=\square$ Answer:	

Example	Notes
3. Evaluate the algebraic expression $3m-5n$ for $m=-4$ and $n=2$. Replace the variables with the given values, and simplify using order of operations. $3(\square)-5(\square)$ $\square-\square=\square$ Answer:	
4. Translate "the difference of 4 and a number" from words to symbols. Use n as the variable. Difference means ____________. Subtract in the order the numbers are stated unless the wording in the expression states otherwise. $\square-\square$ Answer:	

Helpful Hints

Algebraic expressions contain one or more terms and one or more variables.

In an expression, terms are separated by "+" and "−" signs.

Concept Check

1. How many terms in the expression $\frac{1}{2}+2p$? Is it an algebraic expression?

Practice

Evaluate the expression for the given values of the variables.

2. $-6(x-y)$ for $x=-3$ and $y=4$

3. $8m+4q-9$ for $m=5$ and $q=-7$

Identify the terms, coefficients, and variables in each expression.

4. $9-3x+w$

5. $0.5a+(-7b)$

Name: ______________________________ Date: ________________
Instructor: ____________________________ Section: ______________

Introduction to Algebra
12.2 Like Terms

Vocabulary
coefficient • like terms • combining like terms • term

1. Terms that have the same variables raised to the same powers are ____________.

Step-by-Step Video Notes
Watch the Step-by-Step Video lesson and complete the examples below.

Example	Notes
1. Circle the like terms in the list. $-5st,\ st,\ 7t,\ 18st,\ 9s$ Like terms have the same variables raised to the same powers. In this case, the like terms all contain the same variables, $\square$. Answer:	
2. Combine the like terms. $13a-4a+6a$ All three terms are like terms. You can combine (add or subtract) from left to right. $13a-4a=\square \qquad \square+6a=\square$ Answer:	
3. Simplify $6c+3c-13c+c$ by combining like terms. Combine (add or subtract) from left to right. Any time you have a single variable, the coefficient is 1. So, $c=1c$. $6c+3c-13c=\square \qquad \square+c=\square$ Answer:	

Example	Notes
4. Simplify $2+(-3y)-4+6y$ (combine like terms). Remember that like terms can be numbers without variables too. Combine the terms with the variable y. $\square + \square = \square$ Combine the numerical terms. $\square + \square = \square$ Answer:	

Helpful Hints
Like terms must have exactly the same variables and powers. Like terms can be combined, by adding and subtracting their coefficients. You cannot combine unlike terms.

The coefficient of a single variable is 1. For example, $x = 1x$.

Numbers, terms without variables, are also like terms.

Concept Check
1. Are y^2 and $2y$ like terms? Explain.

Practice
Simplify (combine like terms).

2. $-6x+3x+5x$

3. $8a+3a-a$

4. $4w-5z+7w+6$

5. $\frac{3}{8}x+\frac{4}{5}y-\frac{1}{8}x+\frac{1}{5}y-9$

Name: ______________________________ Date: ________________
Instructor: ___________________________ Section: ______________

Introduction to Algebra
12.3 Distributing

Vocabulary
like terms • order of operations • multiplication • distributive property of multiplication

1. The ____________ states that when multiplying a number by a sum or difference, multiply the number outside the parentheses by each number inside the parentheses.

Step-by-Step Video Notes
Watch the Step-by-Step Video lesson and complete the examples below.

Example	Notes
1. Multiply using the Distributive Property. $7(x-5)$ Multiply the 7 by each term in the parentheses. $7(x)-7(\square)$ Simplify each term. $7x-\square$ Answer:	
2. Multiply using the Distributive Property. $(6x+9)5$ Multiply the 5 by each term in the parentheses. Simplify each term. Answer:	

Example	Notes
3. Multiply using the Distributive Property. $10(3a-b+5c)$ Multiply the 10 by each term in the parentheses. Answer:	
4. Multiply using the Distributive Property. $\frac{1}{2}(4x+8)$ Answer:	

Helpful Hints

The Distributive Property of Multiplication can be represented by $a(b+c)=a(b)+a(c)$.

The Distributive Property can be used when there are more than two terms in the parentheses.

Concept Check

1. Does the expression $4(xy)$ require the Distributive Property to simplify?

Practice

Multiply using the Distributive Property.

2. $4(x+2)$

4. $(7x-4)2$

3. $3(4x+5y-6z)$

5. $\frac{1}{3}(6x+9)$

Name: ______________________________ Date: ________________
Instructor: ___________________________ Section: ______________

Introduction to Algebra
12.4 Simplifying Algebraic Expressions

Vocabulary
term • variable • expression • algebraic expression

1. A(n) ____________ is a collection of one or more terms, with multiple terms separated by "+" or "−" signs.

2. A(n) ____________ is an expression that contains one or more variables.

Step-by-Step Video Notes
Watch the Step-by-Step Video lesson and complete the examples below.

Example	Notes
1. Simplify by using the Distributive Property and then combining like terms $5+2(3x-6)$. Distribute $2(3x-6)$ $2(3x-6)=2(3x)-\square(6)$. Simplify $6x-\square$. Combine like terms $5+6x-\square=6x-\square$. Answer:	
2. Simplify $8(4x+5)+2(3x-1)$. (Use the Distributive Property and then combining like terms.) Distribute $8(4x+5)$. Distribute $2(3x-1)$. Combine like terms. Answer:	

Example	Notes
3. Simplify $5x-(3x+11)$. Distribute $-(3x+11)$ by distributing the "–" as -1. Combine like terms. Answer:	

Helpful Hints

Only like terms can be added and subtracted.

Remember that when a "–" is in front of parentheses, a -1 can be distributed.

Concept Check

1. Why are the parentheses removed before combining like terms when simplifying algebraic expressions?

Practice

Simplify by using the Distributive Property and then combining like terms.

2. $7+3(2x-3)$

4. $6x-(4x+3)$

3. $3(5x+1)+4(2x-7)$

5. $2(y+3)-7(5+3y)$

Name: ______________________________ Date: ______________
Instructor: ___________________________ Section: ______________

Introduction to Algebra
12.5 Equations and Solutions

Vocabulary

equation • algebraic expression • solution of an equation • equivalent equations

1. A(n) _____________ is a mathematical statement that two expressions are equal.

2. A(n) _____________ is the number that, when substituted for the variable, makes the equation true.

Step-by-Step Video Notes

Watch the Step-by-Step Video lesson and complete the examples below.

Example	Notes
1. Identify $4x+3=7$ as an expression or an equation. Is there an equal sign? _____ If there is not, then this is an expression; if there is, then this is an equation. Answer:	
2. Determine if 1 is a solution of $2x-3=1$. Replace the variable x with the value 1 in the equation. $2(\square)-3=1$ Simplify the left side. $2(1)-3=\square$ The right side, 1, is already simplified. Are the two sides equal? Answer:	

Example	Notes
3. Determine if -1 is a solution of $x+5=2(x+3)$. Replace the variable x with the value -1 in the equation. Answer:	
4. Determine if $2x+5=9$, $2x=4$, and $x=2$ are equivalent equations. Since $x=2$, determine if 2 is a solution of $2x+5=9$, by replacing the variable x with the value of 2. Do the same for $2x=4$. Answer:	

Helpful Hints
An equation may have one solution, more than one solution, or no solution.

Concept Check

1. How can you determine if a given number is a solution to an equation?

Practice

Determine if the number given is a solution to the following equations.

2. 5 for $3x-4=10$

3. -4 for $3x=6(x+2)$

Determine if the following are equivalent equations.

4. $2x-1=x+3$, $3x=12$, and $x=4$

5. $x+5=2(x+3)$, $x+3=4$, and $x=1$

Name: ______________________________ Date: ________________
Instructor: ___________________________ Section: ______________

Introduction to Algebra
12.6 Solving Equations by Adding or Subtracting

Vocabulary
addition • reverse operation • solution of an equation • addition property of equality

1. The ____________ states that adding the same number to both sides of an equation does not change the solution of the equation.

Step-by-Step Video Notes
Watch the Step-by-Step Video lesson and complete the examples below.

Example	Notes
1. Solve $x-5=3$ by adding or subtracting. Check your solution. To get the variable x by itself, undo what is being done to the variable. What is being done to the variable? ____________ Undo this by adding 5 to each side of the equation. $x-5+5=3+\square$ Simplify by combining like terms. $x=\square$ Check your solution. Answer:	
2. Solve $x+2=-9$ by adding or subtracting. Check your solution. What must be done to both sides of the equation? ____________ Check your solution. Answer:	

Example	Notes
3. Solve $x-9=-5$ by adding or subtracting. Check your solution. Answer:	
4. Solve $-2.5=x+1.3$ by adding or subtracting. Check your solution. Answer:	

Helpful Hints
The reverse operation of addition is subtraction; the reverse operation of subtraction is addition.

If $a=b$, and c is any number, then $a+c=b+c$

Concept Check

1. Explain how the addition property of equality is used to solve $x+2=1$. How does this differ from how it would be used to solve $x-2=1$?

Practice
Solve the equation by adding or subtracting. Check your solution.

2. $x+6=-3$
3. $x-4=7$
4. $5=x-7$
5. $x-1.6=2.8$

Name: ______________________________ Date: ________________
Instructor: ____________________________ Section: ______________

Introduction to Algebra
12.7 Solving Equations by Multiplying or Dividing

Vocabulary
division • reverse operation • solution of an equation • multiplication property of equality

1. The _____________ states that multiplying both sides of an equation by the same number (except zero) does not change the solution of the equation.

Step-by-Step Video Notes
Watch the Step-by-Step Video lesson and complete the examples below.

Example	Notes
1. Use the Multiplication Property of Equality to solve the equation $3x = 18$. To get the variable x by itself, undo what is being done to the variable. What is being done to the variable? _____________ Undo this by dividing each side of the equation by 3. $3x \div 3 = 18 \div \square$ Simplify. $x = \square$ Check your solution. Answer:	
2. Solve by multiplying or dividing. Check your solution. $\frac{x}{5} = 15$ What must be done to both sides of the equation? _____________ Check your solution. Answer:	

Example	Notes
3. Solve by multiplying or dividing. Check your solution. $\frac{3x}{4}=\frac{2}{3}$ Multiply each side of the equation by the reciprocal to the fraction coefficient of x. Answer:	

Helpful Hints
Multiplication and division are reverse operations.

If $a=b$, and c is any number, then $ac=bc$.

Concept Check
1. Why is an equation solved by division instead of multiplication?

Practice
Solve the equation by multiplying or dividing. Check your solution.

2. $5x=35$

4. $0.2x=-1.2$

3. $\frac{x}{4}=16$

5. $\frac{2x}{3}=\frac{4}{9}$

Name: ______________________________ Date: ______________

Instructor: ____________________________ Section: ______________

Introduction to Algebra
12.8 Solving Equations Using Both Properties of Equality

Vocabulary

equation • addition property of equality • solve • multiplication property of equality

1. The ____________ states that multiplying both sides of an equation by the same number (except zero) does not change the solution of the equation.

Step-by-Step Video Notes

Watch the Step-by-Step Video lesson and complete the examples below.

Example	Notes
1. Solve the following equation. Check your solution. $4x-5=7$ Get the variable term alone on one side of the equation. What must be done to each side of the equation? ____________ $4x-5+\square=7+\square$ Simplify. $4x=\square$ Now get the variable alone on one side of the equation. What must be done to each side of the equation? ____________ $4x\div\square=\square\div 4$ Simplify. $x=\square$ Check your solution. Answer:	

Example	Notes
2. Solve the equation. Check your solution. $\frac{x}{5}+2=-7$ Subtract 2 from each side of the equation. Multiply each side of the equation by $\square$. Answer:	
3. Solve the equation. Check your solution. $-4=1+5x$ Answer:	

Helpful Hints

We follow the order of operations when solving equations. Addition or subtraction is used first in solving equations, followed by multiplication or division.

Concept Check

1. Describe the steps used to solve $3x+5=16$.

Practice

Solve the equation. Check your solution.

2. $2x-3=5$

3. $\frac{x}{3}+4=-2$

4. $7-3x=-8$

5. $1=9-4x$

Name: ______________________________ Date: ________________
Instructor: ___________________________ Section: ______________

Introduction to Algebra
12.9 Solving Equations with Multiple Steps

Vocabulary
equation • addition property of • division • multiplication property of equality

1. The ____________ states that multiplying both sides of an equation by the same number (except zero) does not change the solution of the equation.

Step-by-Step Video Notes
Watch the Step-by-Step Video lesson and complete the examples below.

Example	Notes
1. Solve. $3(x+5)=-6$ Remove the parentheses. $3(x)+\square(5)$ Simplify. $3(x)+\square(5)=3x+\square=-6$ Get the variable term on one side of the equation. Get the variable on one side of the equation. Check your solution. Answer:	

Example	Notes
2. Solve. $6x+7=2x-5$ Get the variable terms on one side of the equation. Get the number terms on the other side of the equation. Answer:	
3. Solve. $5(x-4)+7=3(x+5)$ Answer:	

Helpful Hints

Remember to get the variable terms on one side of the equation and the number terms on the other side of the equation when solving multiple step equations.

Concept Check

1. List the seven steps that may be needed in solving a multiple step equation.

Practice

Solve the equation. Check your solution.

2. $2(x+1)=-4$

3. $7x+1-2x=-9$

4. $5x+8=3x+12$

5. $3(3x-1)-4=4(x+2)$